Aims and Scope

The series "Springer Theses" brings together a selection of the very best Ph.D. theses from around the world and across the physical sciences. Nominated and endorsed by two recognized specialists, each published volume has been selected for its scientific excellence and the high impact of its contents for the pertinent field of research. For greater accessibility to non-specialists, the published versions include an extended introduction, as well as a foreword by the student's supervisor explaining the special relevance of the work for the field. As a whole, the series will provide a valuable resource both for newcomers to the research fields described, and for other scientists seeking detailed background information on special questions. Finally, it provides an accredited documentation of the valuable contributions made by today's younger generation of scientists.

Theses are accepted into the series by invited nomination only and must fulfill all of the following criteria

- They must be written in good English.
- The topic should fall within the confines of Chemistry, Physics and related interdisciplinary fields such as Materials, Nanoscience, Chemical Engineering, Complex Systems and Biophysics.
- The work reported in the thesis must represent a significant scientific advance. If the thesis includes previously published material, permission to reproduce this must be gained from the respective copyright holder.
- They must have been examined and passed during the 12 months prior to nomination.
- Each thesis should include a foreword by the supervisor outlining the significance of its content.
- The theses should have a clearly defined structure including an introduction accessible to scientists not expert in that particular field.

Springer Theses

Recognizing Outstanding Ph.D. Research

For further volumes:
http://www.springer.com/series/8790

Ksenia Guseva

Formation and Cooperative Behaviour of Protein Complexes on the Cell Membrane

Doctoral Thesis accepted by
Institute of Complex Systems and Mathematical
Biology of the University of Aberdeen, UK

 Springer

Author
Dr. Ksenia Guseva
Department of Physics
Institute of Complex Systems
 and Mathematical Biology
University of Aberdeen
Meston Building, Aberdeen
AB24 5EE
UK
e-mail: ksenia.guseva@uni-oldenburg.de

Supervisor
Dr. Alessandro P. S. de Moura
Department of Physics
Institute of Complex Systems
 and Mathematical Biology
University of Aberdeen
Meston Building, Aberdeen
AB24 5EE
UK
e-mail: a.moura@abdn.ac.uk

ISSN 2190-5053
ISBN 978-3-642-23987-8
DOI 10.1007/978-3-642-23988-5
Springer Heidelberg Dordrecht London New York

e-ISSN 2190-5061
e-ISBN 978-3-642-23988-5

Library of Congress Control Number: 2011937437

Cover design: eStudio Calamar, Berlin/Figueres

Printed on acid-free paper

Springer is part of Springer Science+Business Media (www.springer.com)

Supervisor's Foreword

The work described here started as part of Ksenia Guseva's Ph.D. project on systems biology of microorganisms, which aimed at describing the homeostasis and metabolisms of bacteria, in particular *E. coli*. It developed into a study of various mechanisms of self-organization which can take place on the cell membrane. The path that was chosen borrowed heavily from Physics: our approach consisted in the careful simplification of the rich original system, and the formulation of an appropriate theoretical description, which enables the extraction of the most fundamental properties of the system, which is not heavily influenced by the neglected details. This approach not only resulted in the answer of important biological questions, but also served as contribution to the larger field of complex systems. To underline the significance of the analysis, we systematically provided particular case studies and arrived at important conclusions for these specific biological systems.

In bacteria the membrane has a very important role, since it controls the movement of substances in and out of the cell. It is composed of many interacting elements, which react to conditions external to the cell, and self-organize in order to achieve a functional purpose. Besides this functional organization, these elements are composed of smaller subunits, which are produced in large numbers, and undergo an intricate process of self-assembly. These two processes are the main subject of this work, which focuses on protein oligomerization and spatial organization on the membrane surface.

The first part of the work deals with protein oligomerization, which is the process of formation of protein complexes which are composed by several subunits. These are produced independently and must join together to become functional. The mechanism of assembly is therefore the most important part of the process. It usually occurs after the subunits are inserted in the membrane and involves their diffusion rates and the interaction strength between them. At a first approximation, this process can be mapped onto an aggregation-fragmentation dynamics, which keeps track of the number of complexes of each possible size in the system in time, according to a given mutual interaction. By imposing a limit on the size of the complex formed—due to geometrical constraints—we obtain a suitable model from which the efficiency of the assembly process can be investigated. From this analysis it was possible to conclude that there is an optimum range of fragmentation rates that increases the efficiency. This range contains the minimum fragmentation rate values that are able to influence the process in the lifetime of the bacteria cell.

The second part focuses on how the function of proteins on the membrane depends on their relative spatial organization. One of the most important types of interactions between proteins is the force resulting from the deformation of the elastic membrane. This force can reorganise the proteins on the membrane surface, and directly influence their function, as in the case of some membrane channels, which are activated by tension (called *mechanosensitive*). These channels change their state if the global tension on the membrane changes, or if the local neighbourhood is deformed, for instance, by the presence of another channel. Thus, the spatial self-organization of these proteins will directly dictate their functional response. We proposed a large-scale model of this process, which incorporates the two possible states of the channels, and their mutual interaction. By analysing the transition from the globally homogeneous and non-homogeneous states, as a function of membrane tension and channel density, we were able to show that the global behaviour is strikingly different from that of isolated channels, and the spontaneous agglomeration leads to both lower activation thresholds and longer response times.

The third and last part describes a particular case where the protein function is directly connected to its assembly process: The proteins are formed to perform a certain task and after it is completed, the complex fragments and the subunits are released back into their free form. We illustrate this process with the formation of large membrane pores in bacteria, which are present on the membrane for limited time and then are disassociated. Several of those pores can be produced at the same time. Therefore the formation of each one of the pores depends on the number of free subunits on the membrane and consequently on the number of other pores that are being formed at the same time. By describing the dynamics of formation and fragmentation by a system of differential equations, we analysed the maximum rate of pore formation depending on their size.

The analysis of the three processes considered in this work serve as very good examples on how a simplifying approach can lead to very central conclusions. By considering the most fundamental ingredients which regulate the behaviour of a system of many interacting elements—and thereby disregarding a great deal of details which would make the analysis intractable if taken simultaneously into account—it is possible to identify the most interesting emerging properties.

Aberdeen, July 2011   Alessandro P. S. de Moura

Preface

In this work we analyse aspects of dynamics and organization of biological membranes from a physical prospective. We provide an analysis of the process of self-assembly and spatial organization of membrane proteins. We illustrate the analysis by considering a channel activated by membrane tension called mechanosensitive channels (MS), in *E. coli* and the twin arginine translocation system (Tat).

We analyse the mechanism of formation of oligomeric protein complexes formed by identical subunits. By derivation of a mathematical approach based on Smoluchowski coagulation equation, we study the efficiency of the process of complex formation, taking into account both irreversible aggregation, as well as fragmentation. We find that a small fragmentation rate increases the efficiency of the formation process, however if the fragmentation rate vanishes the irreversible process is very inefficient.

Our second aim is to determine how the spatial organization can affect the function of channels, which are regulated by elastic forces. We map these short-range interactions into a discretized system, from which we obtain the spatial distribution of the channels and its effect on the gating dynamics. We find that organized channels activate at lower membrane tensions, but possess a delay in the reaction time.

In the last part we determine how the formation of transient pores on the membrane depends on the dynamics of its assembly process. We analyse the pores formed by the Tat complex, which is responsible for protein transport through the membrane. This system functions by polimerization in response to a signal of transport demand from a protein in the cell cytoplasm. The direct correlation of the size of the assembled pore and the size of the protein determines the speed of the translocation process. Using a differential equation approach we obtain that the flux of a given protein depends quadratically on its size.

Dr. Ksenia Guseva

Acknowledgments

First of all I would like to thank my primary supervisor Alessandro Moura. I feel very grateful to him for offering me support, guidance and freedom to create and develop my own ideas and approaches. Also I am very grateful to Prof. Celso Grebogi for his supervision and encouragement. At last but not least I would like to thank my supervisor from the institute of medical sciences Prof. Ian Booth. During the period of my candidature I was fortunate to have collaborated with Dr. Marco Thiel, and I would like to thank him for the fruitful discussions.

I want to thank the Biotechnology and Biological Sciences Research Council, the College of Physical Sciences and the Institute of Medical Sciences of the University of Aberdeen for providing financial support. Also I am very happy for being part of the Institute of Complex Systems and Mathematical Biology and would like to thank all its members. I would like to thank all the very good friends that I had a chance to meet here in Aberdeen: Luca Ciandrini, Julia Slipanchuk, Morgiane Richards, Tao You, Pawel Czaja, Christian Rodrigues, Camila de Alemeida, Komalapriya Chandrasekaran, and Paulo Pinto.

I would like also to deeply thank my family: my mother, my father and my dearest brother Igor. Their examples of persistence, work and ability of happily face life challenges will always reflect throughout my life.

At last I would like to thank Tiago Peixoto, for his love, friendship and for having a lot of patience with me. For never getting complacent, for always being honest with his critics and encourage me to improve. His support was crucial trough all the period of my candidature.

Aberdeen, January 2011   Dr. Ksenia Guseva

Contents

1 Introduction 1
   References 3

**2 The Role of Fragmentation on the Formation
of Homomeric Protein Complexes**......................... 5
   2.1   Introduction...................................... 5
   2.2   Homomeric Protein Complexes 6
        2.2.1   Homomeric Membrane Proteins 8
        2.2.2   Interactions Responsible for the Formation
                of Quaternary Structures....................... 9
        2.2.3   Diffusion Coefficient of Membrane Proteins......... 10
   2.3   The First Passage Time Processes: An Estimation
         of the Aggregation Time Scale 11
        2.3.1   The First Passage Time on a Sphere 12
   2.4   The Smoluchowski Coagulation Equation................ 13
        2.4.1   System with Fragmentation..................... 15
   2.5   The Efficiency of Formation of Protein Complexes 16
        2.5.1   Steady State Size Distribution with
                Irreversible Dynamics 18
        2.5.2   Role of the Fragmentation 20
   2.6   Discussion.. 28
   References 29

**3 Collective Response of Self-Organised Clusters
of Mechanosensitive Channels** 31
   3.1   Introduction...................................... 31
   3.2   Mechanosensitive Channels........................... 32
        3.2.1   Prokaryotic Mechanosensitive Channels 33
        3.2.2   Eukaryotic Mechanosensitive Channels 35

3.3   Experimental Techniques Employed in the Study
       of Mechanosensitive Channels.   36
       3.3.1   Measurement of the Channel Gating and Sensitivity
               to Tension   37
       3.3.2   Spatial Localization of Mechanosensitive Channels   38
3.4   Evidences of Clustering of Mechanosensitive Channels   39
3.5   Individual Channel Gating   39
3.6   Interactions Between Membrane Inclusions.   40
       3.6.1   Direct Protein–Protein Interactions.   41
       3.6.2   Membrane-Mediated Protein–Protein Interactions   42
3.7   Model of the Cooperative Gating of the
       Mechanosensitive Channels.   46
3.8   Dynamics of Agglomeration   47
       3.8.1   The Lattice Gas Phase Diagram   49
       3.8.2   Conditions for Mechanosensitive
               Channel Agglomeration   51
3.9   Dynamics of Gating.   51
       3.9.1   Equilibrium Properties   54
       3.9.2   Dynamics of Escape From the Metastable State   59
       3.9.3   The Transition From Closed to Open Conformations . . .   59
       3.9.4   The Transition From Open to Closed Conformations . . .   61
       3.9.5   Classical Nucleation Theory and the Delay
               of the Channel Response   62
3.10   Discussion.   64
References   65

4   Assembly and Fragmentation of Tat Pores   69
    4.1   Introduction.   69
    4.2   Tat Protein Transport System   69
    4.3   The Theory of TatA Assembly Process.   71
           4.3.1   The Assembly of a Single Ring.   72
    4.4   Assembly of Multiple Rings   73
           4.4.1   Translocation of Only One Type of Protein   74
           4.4.2   Translocation of Protein of Distinct Sizes.   74
           4.4.3   Condition for the Existence of a Steady State   76
    4.5   Discussion.   77
    References   78

5   Conclusion.   79

Chapter 1
Introduction

The attempt to establish fundamental laws which govern the large-scale behaviour of living organisms is still an ongoing challenge being undertaken by the scientific community. Although our qualitative knowledge and the amount of available data are progressively increasing, there is still a perceived need for improved theoretical and quantitative approaches. The identification of fundamental laws governing the behaviour of atoms and molecules, which is the core of modern physics, followed with the ambition to identify similar laws for living organisms. The usage of the atom-molecular approach in treating living systems dates to a lecture given by Schrödinger (which later became a book [1]), entitled "What is life? The physical aspects of a living cell". This work tackled many questions from life sciences which could be answered directly from established physical laws. Schrödinger's attempt to define life as an aperiodic crystal was followed by several, more elaborate alternative definitions of life. Although it is still not possible to define the fundamental characteristic that distinguishes living organisms from other systems [2, 3], it is well accepted that life is an emergent phenomenon, i.e. its global properties are a direct result of the interactions of many simpler constituents, in a way which is not possible to derive from the complete understanding of its constituents in isolation. This feature leads to a non-reductionist approach in life sciences, which is reflected in the relatively new fields of systems biology and synthetic biology (which has a more practical, engineering-like focus). This general approach is also what characterizes the study of "complex systems" in physics.

One of the most central topics in the understanding of emergence and self-organisation in living organisms is the study of biological membranes [4]. The membrane functions as a topological barrier defining the cell entity, and is also responsible for a controlled exchange of substances with the environment. It is one of the most common features, shared between almost all living organisms. The intricate organization of biological membranes, which results in a precise execution of a series of vital process, fascinates scientists of diverse fields. Its complex structure is the result of the biochemical and biophysical activities of its constituents. The self-organisation of the membrane can be divided in three levels: (1) The assembly of lipids into

K. Guseva, *Formation and Cooperative Behaviour of Protein Complexes on the Cell Membrane*, Springer Theses, DOI: 10.1007/978-3-642-23988-5_1, © Springer-Verlag Berlin Heidelberg 2012

bilayers; (2) The assembly of polipeptides into protein complexes; (3) The formation of specialized membranes domains, composed of many mutually-interacting protein complexes.

The self-organisation of lipids to form a bilayer is a vastly explored subject in the field; however the other organisational levels are relatively recent subjects of research, and they are the focus of this particular work. This analysis is of vital importance, since most of the function exerted by the membrane is achieved through proteins, which can constitute up to 75% of the total mass of the membrane. The main focus of this work is to identify emerging behaviours of membrane proteins, which arise of simple interactions among them. We focus on two different processes: (1) The assembly of protein complexes; and (2) the cooperative organization of membrane channels. We employ principles from statistical mechanics to construct theoretical models which are capable of elucidating the most essential properties of the systems, without introducing unnecessary complications.

This thesis is divided in three chapters. The first one describes the assembly of homomeric protein complexes. These complexes are formed by identical subunits, and are very common among living organisms [5]. These subunits spontaneously self-organize during the formation of a complete complex, by means of free diffusion and short-range attractive forces. Using a mean-field approach we describe the assembly process by a master equation, based on the Schmoluchowski coagulation equation [6]. This allows for a dynamical description of the growth of subunit aggregates, and for the identification of the most relevant properties which are responsible for the production of aggregates of the complete size. Our main focus is the characterization of the efficiency of the assembly process, as a function of the fragmentation rate and final complex size. The strength of the interactions among the subunits defines the fragmentation rate, which is a central parameter in our analysis. Our objective is to identify possible ways biological organisms adapted to the need of fast formation of a the required number of complexes.

In the second chapter we move to the cooperative behaviour of formed complexes on the cell membrane. We investigate how spatial reorganization leads to an interplay between individual responses and group behaviour of channels in reaction to changes in the environment. As a case study, we consider a system of bacterial mechanosensitive channels, which represents a minimalistic paradigm of functional self-organisation of proteins on the membrane. These are special channels activated by membrane tension, which prevent bacteria death in a situation of osmotic shock [7]. The channels are capable of changing their conformation from an open to a closed state, according to the membrane deformation in their surroundings. This property leads to an interaction between neighboring channels which can affect their function. Based on a detailed study of the nature and strength of these elastic interactions, we formulate a lattice-based coarse-grained model which is capable of fully describing the emergent properties of the system. We focus on the formation of self-organized protein clusters (high density regions), and their cooperative gating response to osmotic shock. We outline fundamental differences between the cooperative behaviour of clustered channels and isolated ones, and establish their relevance to the physiology of the cell [8].

In the third chapter we return to the assembly process on the membrane. However, at this point we are interested in the dynamics of formation of transient pores that allow protein translocation trough the membrane. We study the role of subunit aggregation in the function of the Tat system [9]. The Tat system consists of subunits that diffuse freely on the membrane and aggregate in response to a translocation request signal, given by a protein inside the cell. The assembly consists in ring formation by continuous monomer incorporation, which is always followed by a complete dissociation and pore sealing. Using a system of differential equations we describe the number of free subunits on the membrane and the number of assembled rings. These quantities depend on the demand rate for protein transport. We analytically obtain the relevant dynamical properties of the system, which relate the dynamics of assembly to the flux of exported proteins through the membrane.

In all three chapters we focus on the relevant time scales. In the first and the last part it is important to consider the time of bacteria replication, since complex formation and protein transport are tightly linked to the cell growth. In both situations, however, the speed of the process is limited not only by the availability of primary resources but also by dynamical process itself. In the second chapter we also analyse the response time of the channels to environmental changes. We observe that increase in number of channels and consequent appearance of cooperative behaviour allow a higher sensitivity to a lower range of tensions, but this comes at the expense of a delayed response time of the whole system.

References

1. Schrödinger, E.: What is life: the physical aspect of the living cell; with, mind and matter and autobiographical sketches. Cambridge University Press, Cambridge (1992)
2. Ivanitskii, G.R.: 21st century: what is life from the perspective of physics? Physics-Uspekhi **53**(4), 327–356 (2010)
3. Kaneko, K.: Life: An Introduction to Complex Systems Biology, 1st edn. Springer, Berlin (2006)
4. Weiss, T.F.: Cellular Biophysics, vol. 1: Transport, The MIT Press, Cambridge (1996)
5. Levy, E.D., Erba, E.B., Robinson, C.V., Teichmann, S.A.: Assembly reflects evolution of protein complexes. Nature **453**(7199), 1262–1265 (2008)
6. Krapivsky, P.L., Redner, S., Ben-Naim, E.: A Kinetic View of Statistical Physics. Cambridge University Press, Cambridge (2010)
7. Booth, I.R., Edwards, M.D., Black, S., Schumann, U., Miller, S.: Mechanosensitive channels in bacteria: signs of closure?. Nat. Rev. Microbiol. **5**(6), 431–440 (2007)
8. Guseva, K., Thiel, M., Booth, I., Miller, S., Grebogi, C., de Moura, A.: Collective response of self-organized clusters of mechanosensitive channels. Phys. Rev. E **83**, 020901(R) (2011)
9. Gohlke, U., Pullan, L., McDevitt, C.A., Porcelli, I., de Leeuw, E., Palmer, T., Saibil, H.R., Berks, B.C.: The TatA component of the twin-arginine protein transport system forms channel complexes of variable diameter. Proc. Nat. Acad. Sci. U. S. A. **102**(30), 10482–10486 (2005)

Chapter 2
The Role of Fragmentation on the Formation of Homomeric Protein Complexes

2.1 Introduction

The synthesis of larger structures by the assembly of several basic building blocks is a general principle in biology. We can see this reflected in the fact that, for example, many proteins are actually complexes formed from several subunits, which spontaneously organize and come together. In a more general way, we could define self-assembly as a process by which ordered aggregates are formed, as a result of the interplay of tightly tuned attractive and repulsive forces in a stochastic environment, with random thermal fluctuations. As such, self-assembly is a widely studied phenomenon in the areas of chemistry, biology and physics [28]. There are many different examples of self-assembly systems, such as crystals, lipid bilayers and many biological structures, such as the aforementioned example of proteins complexes. In this chapter we elaborate a theory of protein assembly, using the Smoluchowski coagulation equation as a simple model. The central objective is to determine the most important parameters which affect the efficiency of the complex formation process. We directly apply the concepts introduced to homomeric membrane complexes of the bacteria *E. coli*.

In chemistry and biology the process of self-assembly is usually reversible and characterized by weak interactions, such as hydrogen bonds. The formed structures are often, therefore, in thermodynamic equilibrium. In this work we analyse the role of reversibility on the formation of proteins complexes, focusing more specifically on role exerted by fragmentation on the final aggregate.

This chapter is organized as follows. At first, in Sect. 2.2 we introduce the biological system we will use as an example, which are the homomeric protein complexes. In Sect. 2.3 we provide a description of the theoretical tools that we employ in this chapter. The important estimations in our study are the time scales of aggregation and of fragmentation. In order to obtain these quantities we consider the relevant stochastic processes, and analyse them using the first-passage time approach. We end the literature review in Sect. 2.4 with an introduction on the Smoluchowski coagulation equation, which is the mean-field approximation in the core of our theory. Then we

K. Guseva, *Formation and Cooperative Behaviour of Protein Complexes on the Cell Membrane*, Springer Theses, DOI: 10.1007/978-3-642-23988-5_2, © Springer-Verlag Berlin Heidelberg 2012

turn in Sect. 2.5.1 to the development of our approach which employs a truncated form of the previously described Smoluchowski equation. We introduce the definition of efficiency of the process, which is simply the ratio of subunits forming the complete protein complex. Then we analyse the irreversible aggregation process, which is highly inefficient when the goal is the formation of large complexes. We follow by extensive studies of the system with fragmentation, which behaves very differently from the irreversible case, with significantly greater efficiency. We have found that the minimisation of the fragmentation rate is a certain way to increase the efficiency of the process, but it cannot vanish such that aggregation becomes irreversible. Furthermore, the contextualisation of our findings for the proposed biological problem shows that the minimisation of the fragmentation rate is not always possible. For the biological system, the total time available to the oligomerisation of proteins is limited by the life time cycle, since a required number of protein complexes should be ready to be transferred to the daughter cells. By imposing such a time constraint, we find the existence of an optimum fragmentation rate, for which the final product is formed with the highest possible efficiency. However for very large complexes, the existence of this optimum fragmentation rate does not imply a large production of final complexes. For this reason we speculate that for a biological system this can consist in an additional evolutionary pressure to keep the complexes small, and to tune the fragmentation rate to correspond to the optimum peaks in efficiency.

2.2 Homomeric Protein Complexes

The past few decades provided a large amount of knowledge about the structure of many proteins. The construction of large data banks of such crystal structures, like protein data bank (PDB)[1] or PISA[2] allowed comparative and evolutionary studies [7, 11, 12, 19]. These analyses characterized a large number of proteins as being composed by several distinct units combined into complexes. They are produced as separate polypeptide chains and assembled into protein complexes as diverse as enzymes, ion channels, receptors, chaperones and transcription factors. These complexes can be either composed by identical subunits, called homo-oligomers (homomers) or they can be formed by distinct interacting parts called hetero-oligomers [1]. As we have described in the Introduction of this chapter we are interested in the process of self-assembly of the first type: the homomeric proteins.

Because these complexes are highly abundant in nature it is speculated that they present evolutionary advantages. Possibly, the most important benefit is that the formation of complexes permits the production of large structures without increasing the genome size. Furthermore, additional advantages include the introduction of an additional level of control, as they can be allosterically regulated; more reliability in transcription, since shorter sequences are more likely to be error-free; the

---

[1]   http://www.pdb.org
[2]   http://www.ebi.ac.uk/msd-srv/prot_int/pistart.html

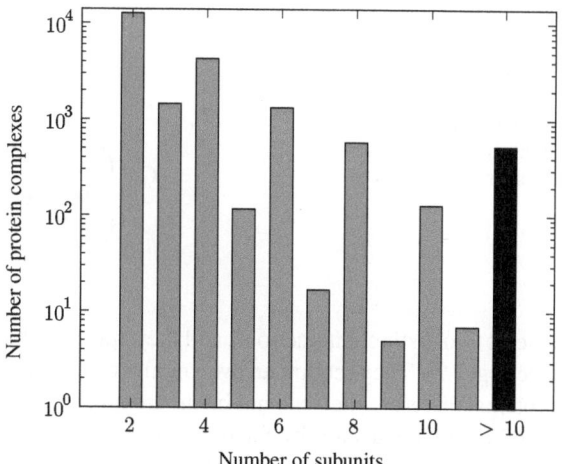

Fig. 2.1 Number of known homomeric proteins complexes (data from PISA)

amplification of evolutionary pressures, since both deleterious and beneficial mutations are more evident; the resistance of larger proteins to degradation and denaturation; the ability to support more complex functions; and finally the larger binding-site specificity of enzymes [1, 7].

The homomeric proteins are particularly abundant among protein complexes and constitute 50−70% of the proteins with resolved quaternary structure[3] [12]. A large fraction of 62% associate only in dimers and the number of proteins decreases exponentially with the number of subunits (see Fig. 2.1). Also it is easy to notice that the number of complexes with odd number of subunits is much smaller than those with an even number (see Fig. 2.1). The preference for the formation of homomeric complexes is due to a strong attraction between identical surfaces. A study from Lukatsky et al. [15] demonstrated theoretically this fact. Their research analysed the energy of interaction between surfaces with random patterns, and they showed numerically that identical random surfaces have always an average minimum energy shifted towards the lower energies, compared with distinct random surfaces.

Another important property of homomeric proteins is their symmetry [1]. Because symmetry confers stability, the homomeric proteins are usually symmetric and have predominantly: cyclical (C_n),[4] dihedral (D_n) or cubic symmetries [3]. The cyclic symmetries involve only one type of interaction, and contain a single axis of rotational symmetry, forming a ring of symmetrically arranged subunits (see Fig. 2.2). Typically, they are involved in functions that require directionality or rotational motion. They usually form tubes and can interact with membranes forming complexes such as ion channels. On the other hand, protein complexes with dihedral symmetry have

---

[3] The primary structure of a protein is its amino acid sequence. The secondary structure the α-helices and β-sheets. The tertiary represents the chain fold. The quaternary structure is the assembly of those folded polypeptide chains.

[4] In this notation C represents the cyclical symmetry of the protein and n the number of subunits that compose this protein. For example: C_6 is a cyclic hexamer.

Fig. 2.2 Examples of the
two possible symmetries of
homomeric complexes: a
dihedral complex of four
subunits, showing two types
of interactions; and cyclic
tetramer and pentamer

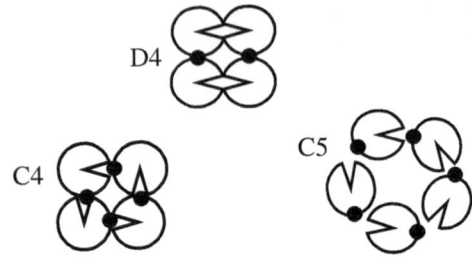

at least two distinct interactions, and they appear to have evolved from cyclic dimers
(see Fig. 2.2). They require an even number of subunits, and are mostly cytoplasmic
proteins, therefore they can not be incorporated in polar biological membranes. The
dihedral complexes are by far more abundant than cyclical ones (it was estimated
that they are at least 10 times more abundant [12]). There are several recent works on
the constrains imposed by the symmetry on the evolution of protein complexes [11].
Here we highlight the work of Villar et al. [25], which focused on the role played
by the relative strengths of the two interactions in dihedral complexes. The authors
used a molecular dynamic approach and showed that there are advantages in the production of assembled proteins using a two step process, which can explain why the
even number of subunits and the dihedral symmetries are more abundant in nature.
However, since membrane proteins have cylindrical symmetry, they could not have
evolved to tune the efficiency of their productions by the strategy described in [25].

As we mentioned previously the cyclic symmetry is predominant in the membrane protein complexes. In the next subsections we show some examples of such
complexes for bacteria and describe their main properties, such as the interaction
strength between subunits and the diffusion coefficient on the cell membrane.

2.2.1 Homomeric Membrane Proteins

The structure of all membrane proteins is constrained by the lipid bilayer. There
are over 200 unique structures of membrane proteins available in the PDB. The
understanding of their function is a challenge in the drug development field since
they are key components for signaling and cell growth control [26]. It has been
reported so far that essential structural themes for the membrane proteins are both
the α-helices and β-barrels. The first ones are predominant in cytoplasmic and cellular
components and the second ones on the outer membrane of bacteria, mitochondria
and chloroplasts [26].

Membrane homomeric proteins can be either channels, pores or receptors with
cyclical symmetry (C_n). As examples we can name: bacterio-rhodopsin (C_3),
AmtB—ammonia transporter *E. coli* (C_3), potassium channels (C_4), acetylcoline
receptors (C_5), FocA—formate transporter *E. coli* (C_5), MscL mechanosensitive

Table 2.1 Stability of the complexes estimated by PISA

	PDB code	Size of the complex	H-bonds	Salt-bridges	BSA
AmtB	1U7G	3 subunits	~4	~3	~17 nm^2
MscL	2OAR	5 subunits	~7	~5	~18 nm^2
FocA	3KCU	5 subunits	~6	~1	~15 nm^2
MscS	2OAU	7 subunits	~25	~8	~26 nm^2

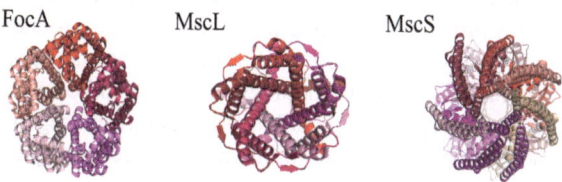

Fig. 2.3 FocA and MscL channels have 5, and MscS has 7 identical subunits

channel of large conductance *E. coli* (C_5), MscS mechanosensitive channel of small conductance *E. coli* (C_7). In Table 2.1 we list some of the proteins from *E. coli* with some of their important characteristics (they all play fundamental roles in the cell survival[5]), for resolved crystal structures see Fig. 2.3.

Since the aim of this work is to study dynamics and the efficiency of their assembly, we will focus on the rate of diffusion and the interaction strength between subunits, which are described in the following two subsections; these two concepts are the ones we will need for modelling the dynamics of the assembly process later on.

2.2.2 Interactions Responsible for the Formation of Quaternary Structures

The quaternary structure of proteins depends on the interactions among the subunits. In general terms, we can summarize all the aspects of the stability of macromolecular complexes and their compatibility of assembly as follows: hydrogen bonds, salt bridges, hydrophobic specificity, interface area; also it is important to take into account the free energy of the complex formation and solvation energy gain. All these aspects are treated in detail with different available software. In this work we use the data available from PISA. Although we do not discuss all of them in detail, in the next part we discuss the buried surface area concept followed by the analysis of hydrogen bonds, which are fundamental for the stability of membrane proteins.

---

[5] The mechanosensitive channels are described in detail in this chapter.

2.2.2.1 Buried Surface Area

The specificity of recognition is estimated from the geometry and chemical properties of the interfaces between subunits. One of the widely used parameters is the interface area, which is the accessible surface area of the subunits that becomes inaccessible to solvent due to protein–protein contacts. This is usually denominated as the buried surface area (BSA) between the macromolecules [9]. It is usual to have the antigen–antibody interaction as a reference measure, which is characterised to be in the range of $12–20\,\mathrm{nm}^2$. Many other protein–protein interfaces are also found to have BSA in this range and have what is considered to be a standard interface area [5, 8].

The BSA is characterised by its topology and the number of hydrogen bonds found on it. Early studies suggested that there is about one hydrogen bond per $2\,\mathrm{nm}^2$ of subunit interface [9]. As we explain in the next subsection, hydrogen bonds are very important for protein stability.

2.2.2.2 Hydrogen Bonds

Hydrogen bonds are known to be one of the main interactions responsible for the stabilization of the secondary structure of proteins and of the formation of protein complexes. In non-polar environments such as biological membranes, these bonds provide enough energy for the establishment of the protein conformation and its tertiary interactions. They are widely considered to be an important force in the membrane environment because of the low dielectric constant of membranes and a lack of competition from water. Indeed, polar residue substitutions are the most common disease-causing mutations in membrane proteins [4]. The strength of the hydrogen bonds can be estimated from the structure of the protein. The strength of the hydrogen bonds in the biological membranes is evaluated to be similar to the strength in vacuum. For $N–H\cdots O$ this corresponds to $\sim 12\,\mathrm{kJ/mol}$. However for an aqueous environment the hydrogen bond energy is lower, around $\sim 6\,\mathrm{kJ/mol}$.

Apart from the interactions which govern the formation of protein complexes, it is necessary to understand how they diffuse on the membrane, if one is to model their efficiency. In the following subsection we provide an estimation of the diffusion coefficient of membrane proteins, based on the Saffman–Delbruck approach [22] and the experimental results of Ramdurai et al. [20].

2.2.3 Diffusion Coefficient of Membrane Proteins

The remaining parameter relevant for the assembly of proteins on the membrane is their diffusion coefficient. More specifically, it is necessary to relate the size of the protein with its diffusion constant on the membrane surface. There are two models used in the literature: in the first one the diffusion constant scales with $1/r$, where r is the radius of the protein; the second model (which is recently gaining credibility)

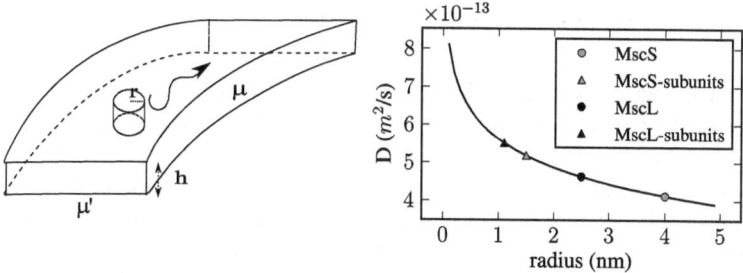

Fig. 2.4 On the *left* a protein of a radius r diffusing on a membrane of thickness h, and viscosity μ. The membrane is surrounded by a liquid of viscosity μ'. On the *right* the diffusion coefficient in an uncrowded membrane as a function of the protein radius according to the Saffman–Delbruck law

considers a logarithmic relation between these two quantities. The relation is given by the Saffman–Delbruck equation [22],

$$D = \frac{k_b T}{4\pi \mu h} \left(\ln\left(\frac{\mu h}{\mu' r}\right) - \gamma \right), \tag{2.1}$$

where k_b is the Boltzmann constant, T is the absolute temperature, h is the thickness of the bilayer, μ is viscosity of the membrane, μ' is viscosity of the outer liquid and γ is the Euler constant. This model was elaborated by Saffman and Delbruck in 1970 to describe the diffusion in highly anisotropic environments such as biological membranes. The diffusion of membrane proteins in *E. coli* was finally analysed in [20], and the results show the validity of the Saffman-Drubruck equation. In Fig. 2.4 we show the prediction, according to this model, for the diffusion of the subunits of mechanosensitive channels, as well as for the assembled functional channel.

2.3 The First Passage Time Processes: An Estimation of the Aggregation Time Scale

It is possible to approach the random walk problem with two types of questions: 1. How far from the starting position x_0 is a particle that moves by Brownian motion after a time t?; 2. Starting from the position x_0 how long it takes for the particle to arrive for the first time at a point x? The second question is very relevant in the context of transport in disordered media, neuron firing, spread of diseases, target search processes, and diffusion-limited reactions. Understanding the first time approach is essential to derive the main theoretical tool which we use in this part, which is the Smoluchowski theory of coagulation. We will start by giving a short introduction to this topic applied to the problem in question. A very good review of first passage time processes can also be found in Sidney Redner's book [21].

2.3.1 The First Passage Time on a Sphere

Since our objective is to describe the diffusion on a surface of a cell we need to describe the first passage time process on a closed surface. We will solve the problem for a particle in Brownian motion on a surface of a sphere of radius R. The particle moves randomly, until it reaches a small absorbent cap, of radius r ($r/R \sim \sin(\delta/2)$), on the north pole of the sphere, in a mean time W. Considering x the particle position, and an infinitesimal time step δt, the mean time is given by the recursion formula

$$W(x) = \frac{1}{2}((W(x + \delta x) + \delta t) + (W(x - \delta x) + \delta t))), \qquad (2.2)$$

which can easily be rewritten as

$$W(x - \delta x) - 2W(x) + W(x + \delta x) = -2\delta t. \qquad (2.3)$$

This turns to

$$DW''(x) = -1 \qquad (2.4)$$

with $D = \frac{1}{2\delta t}$. This equation can be generalised to a Poisson equation for a d-dimensional space,

$$D\nabla^2 W(r) = -1, \qquad (2.5)$$

where ∇^2 is the Laplace–Bertrami operator of the 2-sphere.[6] With this formalism is possible to obtain the average first passage time on a domain with defined boundaries. The problem, therefore, is reduced to a solution of the well-known Poisson equation from electrostatics. Considering a description with spherical coordinates, this system then can be simplified due to its symmetry, since W is independent of the coordinate ϕ,

$$\nabla^2 W(\theta) = \frac{1}{R^2}\left(\frac{d^2 W}{d\theta^2} + \cot\theta\,\frac{dW}{d\theta}\right) = -\frac{1}{D}, \qquad (2.6)$$

with the boundary conditions given by

$$\frac{dW(\pi)}{dt} = 0 \quad W(\delta) = 0. \qquad (2.7)$$

The solution of the equation is[7]

$$W(\theta) = \frac{2R^2}{D}\ln\frac{\sin\theta/2}{\sin\delta/2}. \qquad (2.8)$$

---

[6] This operator corresponds to the Laplace operator on a curved surface.

[7] The maximum time that it takes for the particle to find the cap corresponds to the starting point $\theta = \pi \; W(\pi) = \frac{2R^2}{D}\ln\left(\frac{R}{r}\right)$.

It is also possible to find the average time τ for the particle to reach the absorbing cap from any initial position on the sphere:

$$\langle \tau \rangle = \frac{\int_\delta^\pi W(\theta) \sin(\theta) d\theta}{\int_\delta^\pi \sin(\theta) d\theta} = \frac{2R^2}{D} \ln \left(\frac{R}{r} \right). \tag{2.9}$$

With the theory developed in this section it is possible to estimate the time of formation of the assemblies on the cell surface. It is important to point out that the assembly time will be dependent on the size of the cell and on the diffusion coefficient. The dependence on the radius of the particle, as given by the Saffman–Delbruck model of diffusion discussed previously, is $\tau \sim \ln(R/r) \frac{1}{\ln(1/r)}$. As such, we have a complete approximation of the time scale of the rate of assembly on a spherical surface, which we can use as a parameter in our Smoluchowski-based approach, described in the following sections.

2.4 The Smoluchowski Coagulation Equation

Growth by aggregation is a common natural phenomena found in several different fields, such as astrophysics, colloidal chemistry, polymer science, graph theory and biology [13]. In all these areas, the main process can be described as a dynamics in which complexes of diverse sizes diffuse in the system, and when they approach each other they coalesce in complexes with larger size, which correspond to the sum of the masses of the two originating parts,

$$A_m + A_n \rightarrow A_{m+n}. \tag{2.10}$$

The most studied process in the literature, which we describe in this section, is that of *unbounded* aggregation, where the assembly process proceeds to form infinitely large aggregates. Such situation applies well to systems such as aerosol and liquid droplet coalescence. However, for the study of the assembly of homomeric proteins we use a different approach, which will be described in detail in Sect. 2.5, which imposes an upper bound for the assembly dynamics. This bound for aggregation convey properties to the system which are very distinct from the unbounded variant.

The time evolution of the coagulation process can be described with a master equation approach, where the system is represented by an infinite set of differential equations, which describe the evolution in time of the concentration of each complex size. The main assumptions which need to be made are as follows [10],

1. *The mean-field hypotheses.* It is assumed that there are no spatial correlations of the aggregation reactions, and the particles have a homogeneous distribution in space. Although it is a very strong assumption it is proved to be valid for many physical systems.
2. *Binary collisions.* The reactions happen only between pairs of bodies.

3. *Shape independence.* The aggregation process does not depend on the shape of the aggregate. However it can depend on the aggregate size.

This type of approach to describe the coagulation phenomena was first developed by Marian Smoluchowski in 1917 [23], when he introduced the approach in its continuous form, which can be written as

$$\frac{dn_j}{dt} = \frac{1}{2} \sum_{i=1}^{j-1} K_{i,j-i} n_i n_{j-i} - n_j \sum_{i=1}^{\infty} K_{ij} n_i \qquad (2.11)$$

The members and parameters of these equations are the following:

1. j is the agglomerate size, or units of mass. We define a particle of unit mass as a monomer. Therefore j represents the number of monomers inside a cluster and is also denominated as a j-mer. The number of monomers can vary in the range $[0, \infty]$.
2. n_j is the concentration of particles with size j in the system. It varies in the range $[0, 1]$.
3. K_{ij} is the rate of a specific reaction, called the agglomeration *kernel* and reflects the main dynamical properties of the system. A common assumption is to have symmetric reactions, $K_{ij} = K_{ji}$.
4. M is the total number of units of mass, i.e. monomers, present in the system.

$$M = \sum_{i=0}^{\infty} i n_i(t) \qquad (2.12)$$

It is a conserved parameter during the process. It is possible to verify from Eq. 2.11 that $\frac{dM}{dt} = 0$.

This system was extensively studied for simple kernels such as: $K_{ij} = $ const., $K_{ij} = ij$, and $K_{ij} = i + j$. For most other kernels, an analytical solution has not yet been found. For example, for the kernel representing the rate of collision for particles moving in Brownian motion, with diffusion rates D_u and radius R_u of a complex of size u, which is given by $K_{ij} = (D_i + D_j)(R_i + R_j)$, a general analytical solution has not yet been obtained.

It was shown that the important parameter for this system is not the initial conditions, but only the total mass M. Therefore, it is usual to study the system starting with the simple initial condition

$$n_j(t = 0) = M \delta_{j1} \qquad (2.13)$$

where the system starts only with monomers. The exact solution can be calculated for simple kernels. For the constant kernel, Eq. 2.11, using Eq. 2.12 can be rewritten as

$$\frac{dn_j}{dt} = \frac{1}{2} \sum_{i=1}^{j-1} n_i n_{j-i} - n_j N(t), \qquad (2.14)$$

where $N(t) = \sum_i n_i$ and be solved by a recursive approach, leading to the solution

$$n_j = \frac{t^{j-1}}{(1+t)^{(j+1)}} \tag{2.15}$$

For both the sum and product kernels, an exact solution can be obtained by using generating functions [10]. A very interesting alternative was developed recently by Lushnikov [16], where the system is described as discrete states, with operators of annihilation and creation which act on them. Another alternative is to consider a combined system

$$K_{ij} = \frac{1}{2}(i^\alpha j^\beta + j^\alpha i^\beta) \tag{2.16}$$

where the properties can be analysed in terms of the coefficients α and β [17, 27]. In these systems there is the emergence of very interesting phenomena, such as gelation, which is characterized by the condensation of a non zero fraction of the total mass in a single cluster [10]. The system can be in one of two phases: the gel phase (large cluster) and sol phase (small clusters that disappear with time).

2.4.1 System with Fragmentation

There are many ways to introduce fragmentation in the system, according to the different dynamics that can be in the core of this process, such as random fracture, shattering or homogeneous break-outs. In its simplest form, this process is commonly approached as aggregation running backwards in time (compare with Eq. 2.10) [10]:

$$A_{m+n} \rightarrow A_m + A_n \tag{2.17}$$

The system with aggregation and fragmentation can be represented as

$$\frac{dn_j}{dt} = A(n_1, n_2 \ldots) + \mu' F_j(n_1, n_2 \ldots), \tag{2.18}$$

where $A(n_1, n_2 \ldots)$ is the agglomeration part which can be represented as before by a Smoluchowski coagulation equation. $F_j(n_1, n_2 \ldots)$ is the part that accounts for the fragmentation, and its rate is given by a constant μ'. The first work on coagulation–fragmentation systems was done by Blatz et al. [2] and the fragmentation process was considered as

$$F_j(n_1, n_2 \ldots) = -\frac{n_j}{2}(j-1) + \sum_{i=j+1}^{\infty} n_i \tag{2.19}$$

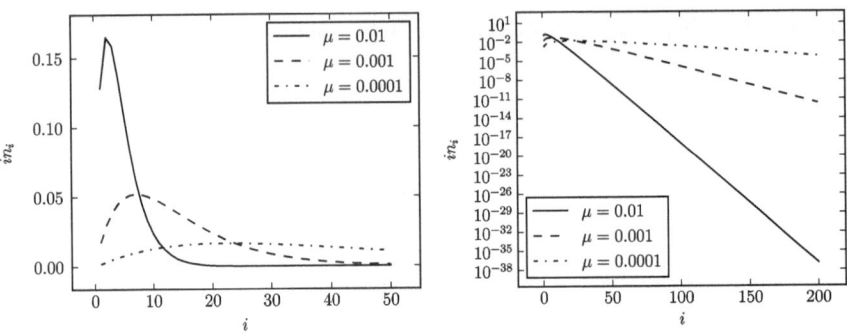

Fig. 2.5 Mass (number of monomers) as a function of size of the cluster i for different fragmentation rates μ

A fragmentation of this type means that larger complexes have more ways to fragment. Here we consider a rescaled version of the parameter $\mu = \frac{\mu'}{K}$, such that we have only one free parameter. The solution of this system is given by [2]

$$n_j = (p)^{j-1}(1-p)^2 \propto e^{(j-1)\ln p} \tag{2.20}$$

where

$$p = \frac{1}{1 + \mu/2 + \sqrt{\mu^2/4 + \mu}} \tag{2.21}$$

An example for the mass distribution in this system is represented in Fig. 2.5. We can see that for a strong fragmentation rate, such as $\mu = 0.01$, the distribution is concentrated in the range of smaller clusters. However for small fragmentation rate ($\mu = 0.0001$), larger aggregates form and the mass is spread on a larger range of aggregate sizes. The decay is always exponential, with exponent $\ln p$.

2.5 The Efficiency of Formation of Protein Complexes

In the following part we present a general theoretical approach for the assembly of protein subunits into oligomeric complexes. Although little is known about the steps of this process, the consensus is that the subunits after production are incorporated in apparently non-correlated locations in the lipid bilayer. They diffuse and by chance approach each other. The short range interactions promote their assembly, and linking bonds between subparts are created. In the case of homomeric proteins, there is a geometrical constrain for the coalescence process. The bounds between subunits are specific, and they establish angles between monomers, which determine the shape of the final structure. This works as a physical barrier for the assembly.

Because we describe an aggregation process, we use the Smoluchowski coagulation approach. As we described previously, the entire reaction dynamics is defined by the coagulation kernel. This approach is equivalent, but largely simpler computationally, to the average calculation of simulations based on individual particle trajectories. Therefore, we highlight that all the numerical solutions presented in this chapter are only numerical solutions of the Smoluchowski coagulation equation. The kernel that we use in the following sections is the constant kernel, which comes from the assumption that every collision leads to an agglomeration event. Therefore, the rate of coagulation reflects directly the dynamics of particle encounters on an approximately spherical surface. As we explained in Sect. 2.3 the first passage time on a sphere is proportional to the radius of the sphere and the diffusion coefficient. The dependence on the radius of the particle, as obtained with the Saffman–Delbruck model of diffusion discussed previously, is $\tau \sim \ln(R/r) \frac{1}{\ln(1/r)}$. Such a logarithmic dependence on the complex size is very weak, especially if the size of the complexes formed do not vary by many orders of magnitude. Therefore it is justifiable to use a constant kernel as a first approximation, since it simplifies the analysis considerably.

We are interested in the situation where the complexes cannot exceed a maximum size N, as is the case of many protein complexes. This will result in a truncated form of the master equation of Eq. 2.11, proposed by McLeod [18].

$$\frac{dn_j}{dt} = \frac{1}{2} \sum_{i=1}^{j-1} K_{i,j-i} n_i n_{j-i} - n_j \sum_{i=1}^{N-j} K_{ij} n_i \qquad (2.22)$$

This is the equation that we will employ in our further studies of protein complexes. Equation 2.11 is its formal limit for $N \to \infty$ [14, 18] and for $M \ll N$. Equation 2.22 has one and only one solution for each set of initial conditions, such that $n_i \geq 0$ (see [18]).

We are particularly interested in the calculation of the ratio of total monomers which are incorporated into the complexes of maximum size. Our objective is to analyse in which conditions this quantity reaches it largest possible value. To quantify this ratio, we use the term *efficiency*, which has the following meaning throughout this chapter.

Definition The efficiency E of formation of particles of maximum possible size N, is the fraction of monomers (n_N) belonging to complexes of size N in the steady state:

$$E = n_N N. \qquad (2.23)$$

We start with the irreversible process and follow with the addition of a fragmentation. We analyse the steady state distribution of the particle sizes.

2.5.1 Steady State Size Distribution with Irreversible Dynamics

In this section we analyse numerically the truncated Eq. 2.22 with a constant kernel, $K_{ij} = 1$. We start the analysis with the study of the steady state $\left(\frac{dn_j}{dt} = 0 \text{ for all } j\right)$. As we can see from the Eq. 2.22 a general characteristic of the system is that all $n_i = 0$ for $i < \frac{N}{2}$. This fact is easy to prove by induction and it does not depend on the initial conditions. On the other hand the values of n_i for $i > \frac{N}{2}$ will all depend on the initial conditions, and any final distribution of n_i, $i > \frac{N}{2}$ is possible. For example, trivially, a set of any initial conditions which has only $n_i > 0$ $\left(\forall\, i > \frac{N}{2}\right)$ will not evolve further in time and will be a fixed point. This is a direct consequence of the size constraint that we impose on the system, since particles of sizes larger then N can not be formed, i.e. a collision between two particles, for which the summed mass exceeds N, does not lead to aggregation.

However we are interested in a specific set of initial condition, where only monomers are present, because this is how we expect to find the protein subunits in the beginning of our process. Therefore we start with mono-disperse initial conditions, $n_j(t = 0) = 1\delta_{j1}$. Subunits assemble into complexes through the evolution of time and the processes stops when subunits of small sizes $\left(i < \frac{N}{2}\right)$ are completely consumed (see Fig. 2.6), since larger complexes with number of particles $i > \frac{N}{2}$ cannot coalesce. An example of a final size distribution can be seen in Fig. 2.7. It is possible to identify three regions in the distribution as follows, for convenience, we assume that N is odd,

$$n_j = 0 \quad \text{for } j < \frac{N}{2} \tag{2.24}$$

$$\frac{dn_j}{dt} = \sum_{i=1}^{N-j} n_i \left(\frac{n_{j-i}}{\delta} - n_j\right) \quad \text{for } \frac{N}{2} < j \le \frac{2N}{3} \tag{2.25}$$

$$\frac{dn_j}{dt} = \sum_{i=1}^{N-j} n_i \left(\frac{n_{j-i}}{\delta} - n_j\right) + \sum_{i=N-j+1}^{i \le N/2} n_i n_{j-i} \quad \text{for } \frac{2N}{3} < j \tag{2.26}$$

where

$$\delta = \begin{cases} 2 & \text{if } i < 2j - N \\ 1 & \text{otherwise.} \end{cases} \tag{2.27}$$

The case for N even can be obtained by including the appropriate terms proportional to $N^2/2$. The two solutions should converge to each other for $N \gg 1$. The particles of size $j < N/2$ are always consumed, and for the larger sizes there are two accumulation behaviours depending on how strong is the effect of the size constraint imposed.

It is also possible to see the effect of the system truncation from the approach introduced by Davies et al. [6]. It is possible to see the evolution in time of the

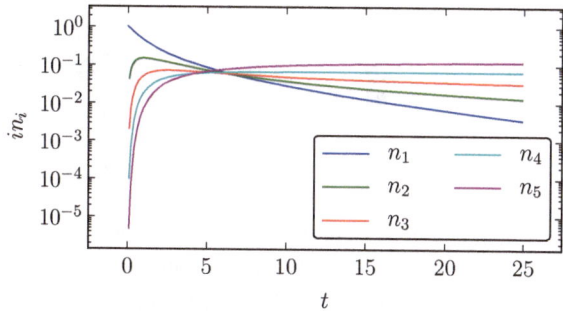

Fig. 2.6 Time evolution of the concentration of j-mers in an aggregation process with a truncation limit $N = 5$, starting with monodisperse initial conditions

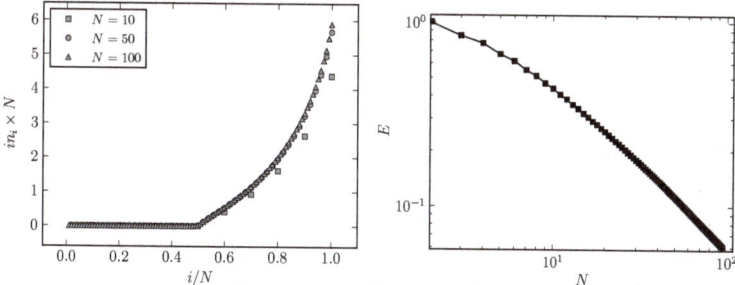

Fig. 2.7 On the left the equilibrium distribution for $N = 100$ starting from monodisperse initial conditions. in_i the total mass of i-mers in the system as a function of particle size i. On the right is the efficiency of the process as a function of truncation size N

process as a flux of mass (monomers) on a lattice with, in our case, N sites. The site r represents the total mass of the particles of size r (rn_r). In this case we introduce J_r which is the mass flux from the cluster at maximum size r to clusters of size larger than r (see Fig. 2.8).

$$J_r = \sum_{j=1}^{r} \sum_{i=r+1-j}^{N-r} j K_{ij} c_j c_i \qquad (2.28)$$

The variation of the mass of particles with size r is therefore

$$\frac{d(rc_r)}{dt} = J_{r-1} - J_r \qquad (2.29)$$

As we can see in Eq. 2.28 the flux J_r decreases as r approaches N. The flux of mass is larger for smaller r and decrease progressively until the site N.

However, the most important thing is to see how this effect is reflected on the efficiency of the process. The efficiency of the process decreases with the increase of

Fig. 2.8 We can also represent the system by a mass flux trough different states of agglomeration using the concept of a flux J_r

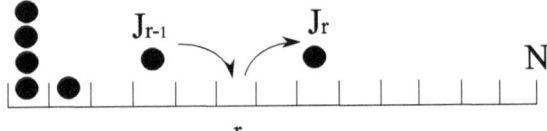

the maximum size N, as can be seen in Fig. 2.7. With the increase of N the number of the intermediate states $N > i > \frac{N}{2}$, in which a subunit can get trapped increases as well. If the total mass of the system is spread through a larger number of intermediates, not many complete complexes are formed at the end of the process.

As we can see, the efficiency of the process is constrained by the assembly dynamics, more specifically by its irreversibility. An alternative therefore is to consider a reversible process, which we do by the introduction of a fragmentation rate in the next section.

2.5.2 Role of the Fragmentation

In this section we introduce a fragmentation rate μ into the process. The system therefore acquires reversible dynamics and in the steady state there is an equilibrium between the assembly of the still incomplete complexes and the fragmentation. The fragmented particles therefore function as a additional inflow of subunits into the system. The dynamics of the system can be represented by Eq. 2.18. However here we use a truncated version of $F_j(n_1, n_2 \ldots)$

$$\frac{dn_j}{dt} = \frac{1}{2} \sum_{i=1}^{j-1} K_{i,j-i} n_i n_{j-i} - n_j \sum_{i=1}^{N-j} K_{ij} n_i - \mu \left(\frac{n_j}{2}(j-1) + \sum_{i=j+1}^{N} n_i \right). \quad (2.30)$$

Similarly to the case of the aggregation kernel, we adopt simple forms of the fragmentation rate. We consider two cases: open and closed chains. For open chains we consider a constant fragmentation rate independent on the size of the complex. In the second case we consider a constant fragmentation rate for all intermediate complexes except for the last one, which is considered to be more stable. This reflects the formation of circular chains of proteins, where the fragmentation of a complete chain implies in breaking not only one, but two links.

2.5.2.1 Uniform Fragmentation Rate (Open Chain)

In this part we analyse the steady state solution of the system with uniform fragmentation. It is possible to find the solution of the steady state, which is given by the expression

$$n_j = \frac{n_1^j}{\mu^{j-1}}, \tag{2.31}$$

for $j > 1$, which can be verified by substitution in Eq. 2.30. The remaining value n_1 can be determined from the relation of conservation of monomers with time

$$M = \sum_{j=1}^{N} \frac{j n_1^j}{\mu^{j-1}}. \tag{2.32}$$

We can see that the distribution depends strongly on the fragmentation rate μ (see Fig. 2.9). In this case we also produce intermediates and the proportion of these intermediates is related to μ. As the fragmentation rate increases, the predominant size of intermediate complexes changes from the maximum size N to smaller values. In order to obtain this dependence on μ more precisely, we differentiate Eq. 2.31 in respect to μ,

$$\frac{dn_j}{d\mu} = \frac{n_1^{j-1}}{\mu^{j-1}} \left(j \frac{dn_1}{d\mu} - (j-1) \frac{n_1}{\mu} \right). \tag{2.33}$$

On the other hand, the differentiation of Eq. 2.31 gives us

$$\frac{dn_1}{d\mu} = \frac{n_1}{\mu} \left(1 - \frac{M}{M_1(\mu, n_1(\mu))} \right), \tag{2.34}$$

where

$$M_1(\mu, n_1(\mu)) = \sum_{i=1}^{N} \frac{i^2 n_1^i}{\mu^{i-1}}, \quad M_1(\mu, n_1(\mu)) = n_1 \frac{dM}{dn_1}. \tag{2.35}$$

Using Eq. 2.33 we obtain

$$\frac{dn_j}{d\mu} = \frac{n_j}{\mu} \left(1 - \frac{jM}{M_1(\mu, n_1)} \right). \tag{2.36}$$

The equation above shows us that $n_N(\mu)$ is a strictly decreasing function, since $NM > M_1$ always holds, as can be seen in Fig. 2.9. This equation also shows that n_1 is always an increasing function with μ, since $M < M_1$. However it is not as easy to obtain the behaviour of other intermediates, n_j with μ. We can calculate the maximum of the function, in other words the value of μ for which n_j is the largest, by solving the polynomial

$$\sum_{i=1}^{N} (i^2 - ji) \frac{n_1^i}{\mu^{i-1}} = 0. \tag{2.37}$$

This analysis shows how different is the behaviour of the system with the introduction of fragmentation. The efficiency of the process increases for $\mu \to 0$, however at the

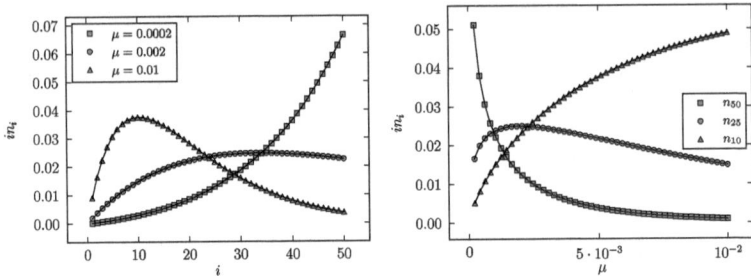

Fig. 2.9 Distribution of mass in$_i$ as a function of particle size i (*left*) and μ (*right*) for $N = 50$. The *dots* are results from numerical solutions of Eq. 2.30, and the *solid lines* are the analytical results given by Eqs. 2.31 and 2.32

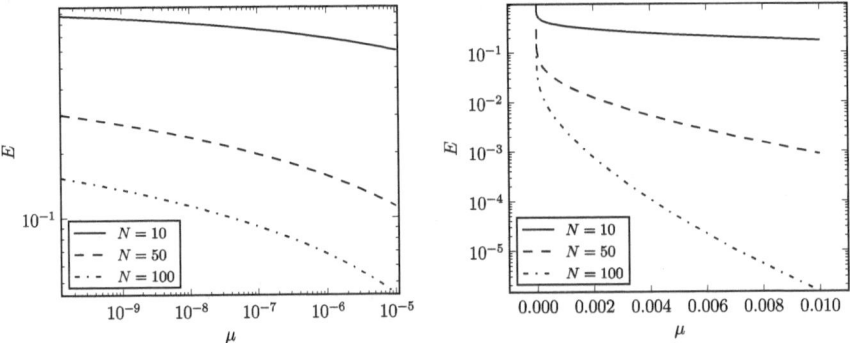

Fig. 2.10 The efficiency E as a function of μ, for different values of N. The graph on the left is close up of the graph on the right side, for small values of μ

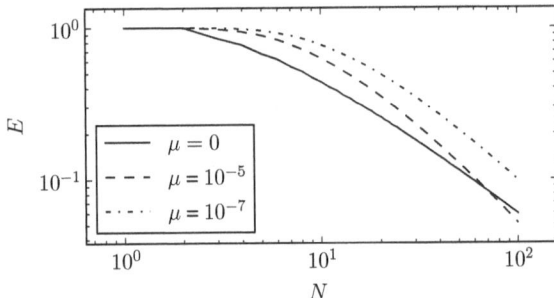

Fig. 2.11 The efficiency of formation of complexes of maximum size as a function of truncation size N, for the system without the fragmentation rate (*solid line*) and for the system with fragmentation (*dashed line*). There are a small increase in the efficiency of the process for small fragmentation rate. However the efficiency still strongly decreases as we increase N equivalently to the case with irreversible dynamics

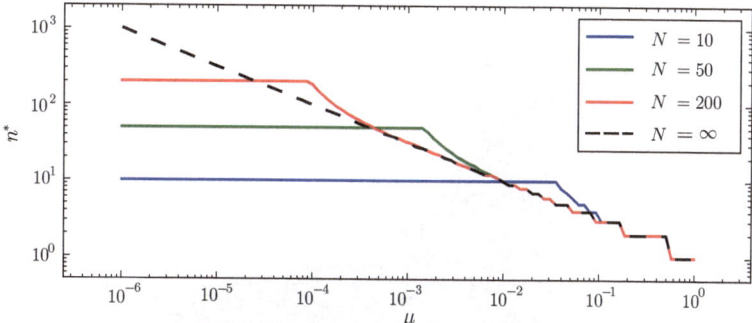

Fig. 2.12 The n^* is the size of particles present in larger concentration in the steady state distribution of the system. The complete complex size is the most formed j-mer for a wide range of μ. The distribution approaches the non-truncated Smoluchowski coagulation equation for large fragmentation rates

point $\mu = 0$ the system is again irreversible and the efficiency drops discontinuously (see Figs. 2.9, 2.10, 2.11).

The behaviour of the system changes in important ways as the value of μ is varied: For small values of μ, the constraint of the maximum size will be the main factor to determine the steady state distribution. However in a system with large values of μ, the constraint stops influencing the dynamics, since the fragmentation does not permit the mass flow in the system to reach the maximum size. In this situation the dynamics is identical to a traditional Smoluchowski coagulation with $N = \infty$. This fact is illustrated in Fig. 2.12, where it is shown the size n^* of the particles that dominates in the distribution of the mass of the steady state, $n^* n_{n^*} \geq j n_j \; \forall j$, as of function of μ. For small values of μ the predominant size is the maximum possible size N. However as the μ increases the fragmentation does not allow the mass flow to distribute significantly further a certain j-mer ($j < N$). In this case the predominant size that captures the larger number of monomers in the steady state is an intermediate size. A similar message is shown in Fig. 2.13, where the system is divided in two regions, as a function of the parameters μ and N: the black region represents the system with the steady state mass belonging predominantly to particles of maximum size ($E > 0.5$), and the white region it is the intermediate particles which dominate ($E < 0.5$).

We now turn to a relevant aspect of aggregation of protein complexes which is the time necessary for their formation. We have obtained so far that smaller fragmentation leads to higher efficiency. However, an aggregation process which relies on a very small fragmentation rate will also tend to be very slow, and in some cases there will be a limitation for the time of aggregation. Such a time limit is present, for example, in a living organism which is expected to duplicate the number of all its arsenal of proteins during a replication cycle. Therefore it is interesting to analyse the assembly product after a given time, in a stage before the system achieves the steady state. The simplest consequence of this limitation is that very small fragmentation rates will

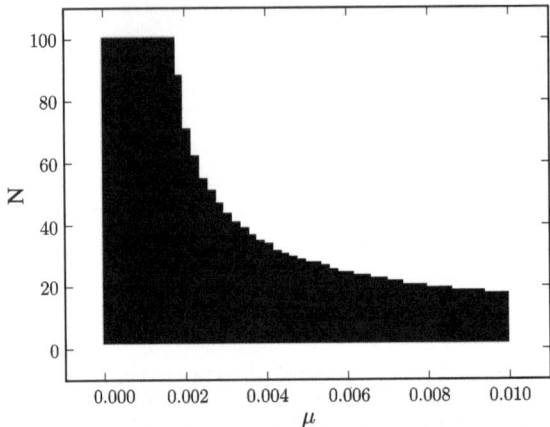

Fig. 2.13 In *black* the region of parameters (N-maximum of the size grow and μ-fragmentation rate) where the aggregates of maximum size N have the larger total mass, $Nn_N > in_i \ \forall i \neq N$

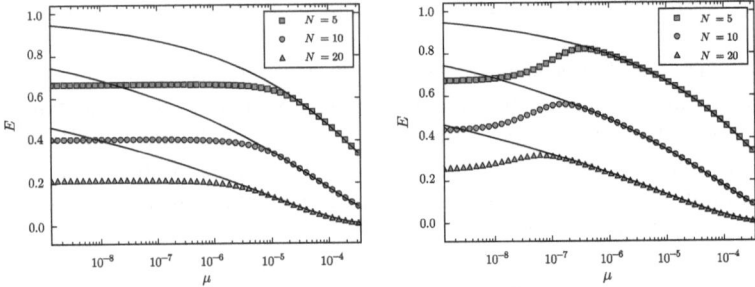

Fig. 2.14 The efficiency E of the assembly process as a function of μ different values of N for a system in a transient state (*points*) and in the steady state (*solid lines*). On the *left* the iteration time is equivalent to 1,000 sec and on the *right* to 10,000 sec

not have any effect. Our observation is that with this limitation there is an optimum fragmentation rate (see Fig. 2.14), and that this optimum decreases with the maximum complex size. For a *E. coli* bacteria the replication time is 20 min. The association kernel can be estimated from the Eq. 2.9 to be $\sim 0.01 \ \text{s}^{-1}$. Figure 2.14 shows how the efficiency of the assembly changes with the protein size for the estimated time. From this estimation we see that very large complexes with this limitation can not be formed efficiently, even if the fragmentation rate is very low.

2.5.2.2 Fragmentation Rate for Rings (Closed Chains)

In this part we consider a system where the complex of maximum size has a higher stability then its intermediates. In this case the monomers assemble in to a circular

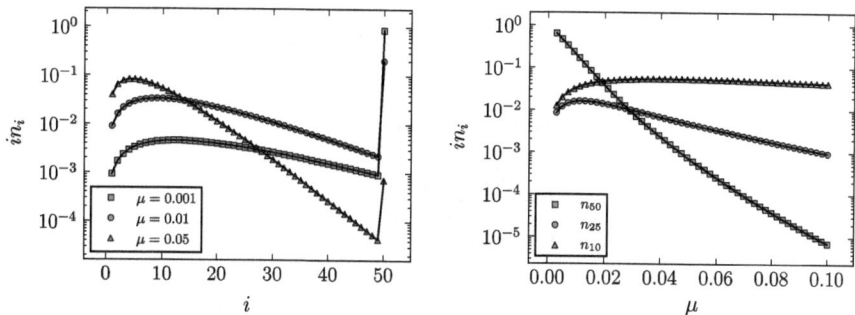

Fig. 2.15 Distribution of particle mass in$_i$ as a function of particle size i, for *closed rings* with $N = 50$. On the *left* we show the final distribution for three fragmentation rates μ. On the *right* we show how the efficiency of formation of i-mers changes with μ, for different values of i

geometry that closes into a ring when the final size complex is achieved. All the intermediate complexes will have a fragmentation rate μ. However due to the circular geometry, two links have to break in order to fragment a complex with maximum size, and therefore the fragmentation rate for it will correspond to μ^2,

$$F_j(n_1, n_2 \ldots) = -\frac{n_j}{2}(j-1) + \sum_{i=j+1}^{N-1} n_i + \mu n_N \quad j \neq N \tag{2.38}$$

$$F_N(n_1, n_2 \ldots) = -\mu \frac{n_N}{2}(N-1) \tag{2.39}$$

Similarly to the case of open chains, we can find the steady state distribution. In the case of $n_j > 0$, the steady state solution is

$$n_j = \frac{n_1^j}{\mu^{j-1}} \quad 1 < j < N, \tag{2.40}$$

$$n_N = \frac{n_1^N}{\mu^N}, \tag{2.41}$$

where n_1 can be determined from the relation of conservation of monomers with time

$$M = \sum_{j=1}^{N-1} \frac{j n_1^j}{\mu^{j-1}} + \frac{n_1^N}{\mu^N}. \tag{2.42}$$

As we can see from the Fig. 2.15 by the decrease in the fragmentation rate of the complete complexes strongly favors their concentration, as expected. As for the previous case we analyse the dependence of n_N on μ.

$$\frac{dn_N}{d\mu} = \frac{Nn_1^{N-1}}{\mu^N} \left(\frac{dn_1}{d\mu} - \frac{n_1}{\mu} \right) \tag{2.43}$$

Also differentiation of the conservation relation gives us:

$$\frac{d}{d\mu} \left(\sum_{j=1}^{N-1} \frac{jn_1^j}{\mu^{j-1}} \right) + N\frac{dn_N}{d\mu} = 0 \tag{2.44}$$

$$\frac{d}{d\mu} \left(\sum_{j=1}^{N-1} \frac{jn_1^j}{\mu^{j-1}} \right) = \sum_{j=1}^{N-1} \frac{j^2 n_1^{j-1}}{\mu^{j-1}} \left(\frac{dn_1}{d\mu} - \frac{n_1}{\mu} \right) + \frac{n_1}{\mu} \sum_{j=1}^{N-1} \frac{n_1^{j-1}}{\mu^{j-1}} \tag{2.45}$$

A substitution in Eq. 2.44 results in:

$$\frac{1}{n_1} \left(\sum_{j=1}^{N-1} \frac{j^2 n_1^j}{\mu^{j-1}} + \frac{N^2 n_1^N}{\mu^N} \right) \left(\frac{dn_1}{d\mu} - \frac{n_1}{\mu} \right) + \frac{1}{\mu} \sum_{j=1}^{N-1} \frac{n_1^j}{\mu^{j-1}} = 0 \tag{2.46}$$

Defining M_1 as:

$$M_1(\mu, n_1(\mu)) = \sum_{i=1}^{N} \frac{i^2 n_1^i}{\mu^{i-1}} + \frac{N^2 n_1^N}{\mu^N}, \quad M_1(\mu, n_1(\mu)) = \frac{1}{n_1} \frac{dM}{dn_1} = 0 \tag{2.47}$$

We end up with the expression

$$\frac{dn_1}{d\mu} = \frac{n_1}{\mu} \left(1 - \frac{(M - Nn_N)}{M_1} \right). \tag{2.48}$$

This expression shows that n_N is always a decreasing function with μ, since

$$\frac{dn_N}{d\mu} = -\frac{MNn_N}{M_1\mu} \left(1 - \frac{Nn_N}{M} \right) < 0. \tag{2.49}$$

The efficiency of this process is higher (Fig. 2.16) compared to the system with an open chain (Fig. 2.11), however it has also the same qualitative behaviour with $\mu \to 0$ (Fig. 2.15). The formation of the complexes with the maximum size is still predominant even for relatively large μ, compared to the open chain system.

As it possible to see from Fig. 2.15, this system has two local maxima, one for the particle of maximum size and another determined by the fragmentation rate. A predominant fraction of the system's mass often corresponds to the particles with maximum size N, until the stage when it ceases to be affected by the truncation completely. At this point the system behaves analogously to the system without truncation (see Fig. 2.17). This result is very similar to the case of an open chain analysed before.

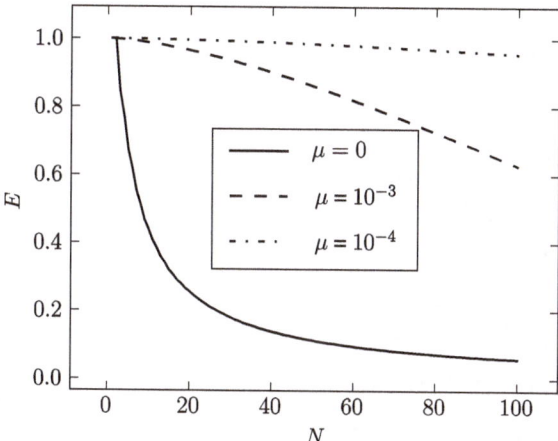

Fig. 2.16 The efficiency of formation of complexes of maximum size as a function of truncation size N, for the system without the fragmentation rate (*solid line*) and for the system with fragmentation (*dashed lines*)

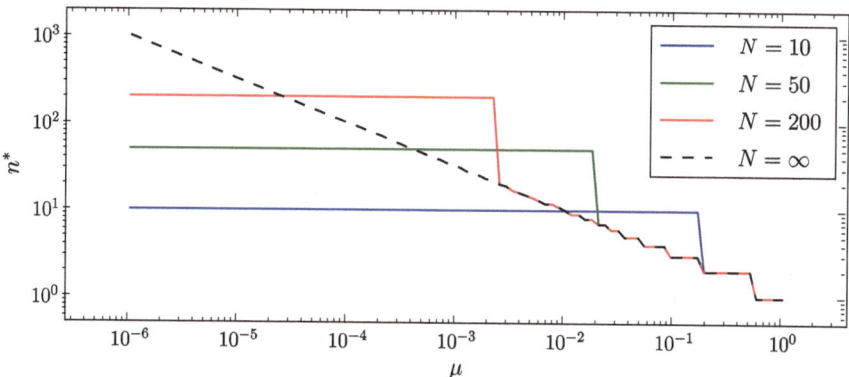

Fig. 2.17 The n^* is the size of particles present in larger concentration in the steady state distribution of the system. The complete complex size is the most formed j-mer for a wide range of μ. The distribution approaches the non-truncated Smoluchowski coagulation equation for large fragmentation rates (*dashed lines*)

It is also possible to analyse this system for a time-limited dynamics, similarly as done in the case for open chains. In this case the system is strongly robust to strong fragmentation rates, however with the imposition of a time limit, lower values of fragmentation also do not lead to higher efficiency, as can be seen in Fig. 2.18. It can be observed that there are optimum values of fragmentation, for which the value of E is maximum. From Fig. 2.18, we can see that the formation of large complexes is strongly constrained by the time limitation. For a time of 1,000 sec, most complexes cannot be formed with maximum efficiency, regardless of the value of μ. For a biological system, this can imply an extra evolutionary pressure to keep

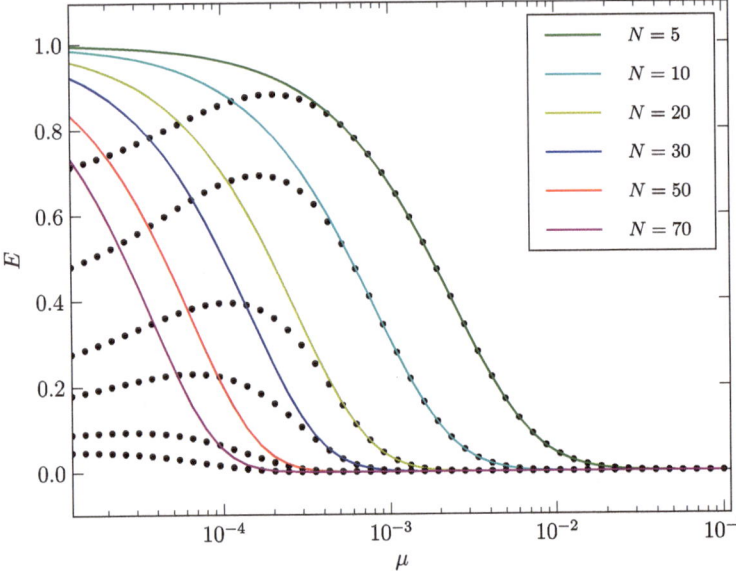

Fig. 2.18 The efficiency E of the assembly process for closed chains as a function of μ different values of N for a system in a transient state after 1,000 sec (*points*) and in the steady state (*solid lines*)

the complexes small, and to tune the fragmentation rate to correspond to the optimum peaks in efficiency.

2.6 Discussion

In this chapter we analysed the formation of protein complexes on the cell membrane with the use of a truncated form of the Smoluchowski coagulation equation. Our aim was to understand the role of fragmentation on the efficiency of the process. In the case of zero fragmentation rate, the irreversibility of the process leads to low efficiency, because the steady state is characterized by a large amount of incomplete intermediates that fail to achieve the maximum size. This is a consequence of the fact that as the small particles are consumed, the growth of the particles of large size is constrained. This effect gets stronger for larger complexes, due to the larger number of intermediate states in which the complexes can be trapped before achieving its final size. The introduction of fragmentation changes this behaviour, because it allows the system to reverse the dynamics with a rate μ. The equilibrium between aggregation and fragmentation results in several possible types of final distributions depending on the parameter μ. Some of these distributions are far more efficient than the irreversible process alone. Furthermore we were able to prove that a dynamic with infinitesimally small but non-zero fragmentation ($\mu \to 0$) is the most efficient, independently on how large is the final size that is necessary.

The application of these results to the formation of homomeric protein complexes suggests possible pressures to which their evolution was subjected: larger complexes (those with a larger number of subunits) must have smaller fragmentation rates, which implies more specific interactions between their subunits, to allow for an efficient assembly process. This requires larger interaction surfaces and a larger number of bonds between subunits. We compare the data presented in Table 2.1, where it is possible to see how the number of hydrogen bonds and BSA increases with the number of subunits that form the complete complex. As more protein structures become available we suggest a systematic continuation of this comparative study.

A decrease of the fragmentation rate necessarily implies a longer time to achieve equilibrium. Therefore the most efficient assembly processes would have to take a longer time. For *E. coli* bacteria the replication time is of approximately 20 min. During this time the bacteria doubles in size and the number of proteins received by the two daughter cells should be identical to the original. Therefore there is an upper bound for the available time for protein formation. For the aggregation kernel equivalent to 0.01 s^{-1}, the fragmentation below $\sim 10^{-5} \text{ s}^{-1}$ will not be relevant for the process. This implies that each organism will have an upper bound for the larger possible complex size that it can form efficiently, and that for each limited time of formation and protein size there is an optimum fragmentation rate.

We propose as a possible continuation of this theoretical study the introduction of distinct aggregation kernels and fragmentation rates. One of the possible alterations is to introduce a diffusion rate which changes with the size of the complex and see how that would alter the dynamics. Also a fragmentation rate that decreases with the complex size or corresponds to certain empirical data would be of interest. Additionally, as an application of the theory, one could investigate the pore forming toxins of bacteria, which also consist of subunits that assemble on the infected membrane, and form holes causing host cell leakage. Since "bacteria toxins are some of the most potent poisons to man" [24], the understanding of their process of assembly can be relevant for pharmaceutical applications.

References

1. Ali, M.H., Imperiali, B.: Protein oligomerization: how and why. Bioorg. Med. Chem. **13**(17), 5013–5020 (2005)
2. Blatz, P.J., Tobolsky, A.V.: Note on the kinetics of systems manifesting simultaneous polymerization-depolymerization phenomena. J. Phys. Chem. **49**(2), 77–80 (1945)
3. Blundell, T.L., Srinivasan, N.: Symmetry, stability, and dynamics of multidomain and multi-component protein systems. Proc. Natl. Acad. Sci. U. S. A. **93**(25), 14243–14248 (1996)
4. Bowie, J.U.: Membrane protein folding: how important are hydrogen bonds? Curr. Opin. Struct. Biol. (2010, in press, corrected proof)
5. Lo Conte, L., Chothia, C., Janin, J.: The atomic structure of protein-protein recognition sites. J. Mol. Biol. **285**(5), 2177–2198 (1999)

6. Davies, S.C.: Self-similar behaviour in the coagulation equations. J. Eng. Math. **36**, 57–88 (1999)
7. Dayhoff, J.E., Shoemaker, B.A., Bryant, S.H., Panchenko, A.R.: Evolution of protein binding modes in homooligomers. J. Mol. Biol. **395**(4), 860–870 (2010)
8. Hwang, H., Pierce, B., Mintseris, J., Janin, J., Weng, Z.: Protein-protein docking benchmark version 3.0. Proteins **73**(3), 705–709 (2008)
9. Janin, J., Bahadur, R.P., Chakrabarti, P.: Protein-protein interaction and quaternary structure. Q. Rev. Biophys. **41**(2), 133–180 (2008)
10. Krapivsky, P.L., Redner, S., Ben-Naim, E.: A Kinetic View of Statistical Physics. Cambridge University Press, Cambridge (2010)
11. Levy, E.D., Pereira-Leal, J.B.: Evolution and dynamics of protein interactions and networks. Curr. Opin. Struct. Biol. **18**(3), 349–357 (2008)
12. Levy, E.D., Erba, E.B., Robinson, C.V.: Assembly reflects evolution of protein complexes. Nature **453**(7199), 1262–1265 (2008)
13. Leyvraz, F.: Scaling theory and exactly solved models in the kinetics of irreversible aggregation. Phys. Rep. **383**(2–3), 95–212 (2003)
14. Leyvraz, F., Tschudi, H.R.: Singularities in the kinetics of coagulation processes. J. Phys. A. Math. Gen. **14**(12), 3389–3405 (1981)
15. Lukatsky, D.B., Zeldovich, K.B., Shakhnovich, E.I.: Statistically enhanced self-attraction of random patterns. Phys. Rev. Lett. **97**(17), 178101 (2006)
16. Lushnikov, A.A.: Exact kinetics of the sol-gel transition. Phys. Rev. E **71**(4), 046129 (2005)
17. Lushnikov, A.A.: Critical behaviour of the particle mass spectra in a family of gelling systems. Phys. Rev. E **76**(1), 011120 (2007)
18. McLeod, J.B.: On an infinite set of non-linear differential equations. Q. J. Math. **13**(1), 119 (1962)
19. Pereira-Leal, J.B., Levy, E.D., Kamp, C., Teichmann, S.A.: Evolution of protein complexes by duplication of homomeric interactions. Genome Biol. **8**(4), R51 (2007)
20. Ramadurai, S., Holt, V.K.A., van den Bogaart, G., Killian, J.A., Poolman, B.: Lateral diffusion of membrane proteins. J. Am. Chem. Soc. **131**(35), 12650–12656 (2009)
21. Redner, S.: A Guide to First-Passage Processes. Cambridge University Press, Cambridge (2001)
22. Saffman, P.G., Delbrück, M.: Brownian motion in biological membranes. Proc. Natl. Acad. Sci. U. S. A. **72**(8), 3111–3113 (1975)
23. von Smoluchowski, M.: Versuch einer mathematischen Theorie der Koagulationskinetik kolloider Lösungen. Z. Phys. Chem. **92**, 124–168 (1917)
24. Tilley, S.J., Saibil, H.R.: The mechanism of pore formation by bacterial toxins. Curr. Opin. Struct. Biol. **16**(2), 230–236 (2006)
25. Villar, G., Wilber, A.W., Williamson, A.J., Thiara, P., Doye, J.P.K., Louis, A.A., Jochum, M.N., Louis, A.C.F., Levy, E.D.: Self-assembly and evolution of homomeric protein complexes. Phys. Rev. Lett. **101**(11), 118106 (2009)
26. Vinothkumar, K.R., Henderson, R.: Structures of membrane proteins. Q. Rev. Biophys. **43**(1), 65–158 (2010)
27. Wattis, J.A.D.: An introduction to mathematical models of coagulation-fragmentation processes: a discrete deterministic mean-field approach. Phys. D Nonlinear Phenom. **222**(1–2), 1–20 (2006)
28. Whitesides, G.M., Boncheva, M.: Beyond molecules: self-assembly of mesoscopic and macroscopic components. Proc. Natl. Acad. Sci. U. S. A. **99**(8), 4769–4774 (2002)

Chapter 3
Collective Response of Self-Organised Clusters of Mechanosensitive Channels

3.1 Introduction

The spatial organization of cellular components can strongly influence the function of biological systems. In this chapter we analyse the consequences of self-organization on the function of channels sensitive to mechanical forces on the cell membrane. These channels have properties that permit them to mutually interact, and enable them to change from an individual to a cooperative behaviour. This characterizes a type of emergent behaviour, where the global properties of the system do not reflect directly the behaviour of its constituents. Emergence and self-organization are widespread phenomena in nature, and are the hallmarks of the area of "complex systems", which spans diverse fields such as nonlinear dynamics, statistical mechanics, network theory, and biology. In this chapter, we focus on an approach based on statistical mechanics, where we map the system of interest to a simplified model, of which an extensive analysis is possible, and which is capable of qualitatively predicting important properties of the original system.

The cooperative behaviour of interacting proteins on the cell membrane is a very current and important topic of investigation [1]. It is believed that many proteins exert their function through global interaction with other proteins of the same type. One paradigmatic example are the Mechanosensitive channels, which are sensitive to mechanical stimuli. These channels are particularly sensitive to their local membrane environment, and will change their conformation accordingly. Neighbouring channels will therefore affect each other via membrane deformations, which, under the right circumstances, may lead to a collective behaviour of several channels. The behaviour of individual channels is well understood, in particular the mechanosensitive channels of *E. coli* [2]. In these bacteria, their main role is to sense the tension of the cytoplasmic membrane, and respond in case of osmotic stress. However these types of channels are present across all realms of life, and are responsible for diverse functions, such as volume regulation, locomotion, and sensory input and signalling. In many of these organisms, mechanosensitive channels occur in very large numbers, and their cooperative behaviour is still not well understood.

K. Guseva, *Formation and Cooperative Behaviour of Protein Complexes on the Cell Membrane*, Springer Theses, DOI: 10.1007/978-3-642-23988-5_3, © Springer-Verlag Berlin Heidelberg 2012

The objective of this chapter is to construct a theory for the self-organization and cooperative behaviour of mechanosensitive channels [3]. We employ a statistical mechanics approach, and describe the system in a simplified manner, where the channels can occupy positions in a discrete lattice, and their conformations are described by spin variables. In this manner we are able to obtain the conditions necessary for channel agglomeration and collective gating, as well as characterize some dynamical features of the global behaviour. We find that the collective behaviour of mechanosensitive channels is drastically different from that of isolated channels, in many biologically relevant parameter regions, which has strong implications for the functioning of the cell. One of our major findings is that channel clustering leads to a lower threshold of channel activation, causing the clustered channels to open for much lower membrane tensions. Furthermore, clustering leads to an increase in the time it takes for the clustered channels to open in response to osmotic shock, which can have drastic consequences for cell survival.

This chapter is organized as follows. In Sect. 3.2 is a general introduction to the field of mechanosensitive (MS) channels. We start it with a short description of the two most well described mechanosensitive channels of bacteria: mechanosensitive channels of large conductance (MscL) and mechanosensitive channels of small conductance (MscS). In Sect. 3.2.2, we follow with a short overview of MS channels in eukaryotes. Section 3.3 describes the experimental techniques used in the area. We start with the patch clamp and the methodology developed by Kung et al. [4] twenty years ago that made possible to obtain extremely large *E. coli* cells to be used with the patch clamp approach. Then we describe the GFP method of labeling of mechanosensitive channels which is used to relate the channel spatial localisation to its activity.

Section 3.6 is a detailed analysis of possible interactions between proteins on the membrane surface. We start with a review of the three possible protein–protein interactions: electric, entropic and elastic. We then focus on the last one, since it is the most important type of force in the context of mechano-transduction. We then quantify this type of interaction between MS of *E. coli* using their crystal structure.

In Sect. 3.8 we introduce the simplified model of channel aggregation which takes into account the nearest–neighbor interactions on a two-dimensional discrete lattice, and we obtain the conditions necessary for channel agglomeration, based on the interaction strength and channel density. In Sect. 3.9 we map the gating dynamics of interacting channels into an Ising-like model of interacting spins, with the presence of a spatially-correlated field, which depends on the spatial distribution of the channels.

3.2 Mechanosensitive Channels

It is useful to classify ion channels in three major groups according to their gating mechanism: the voltage-gated, the ligand (receptor)-gated and the mechanically-gated. In this study we are particularly interested in the last group, which are called MS and are found in the all three domains of life [5, 6]. Although they do not have the same evolutionary origin, they all are characterised by being able to respond to mechanical stimuli. These stimuli can be intracellular (transmitted by the cytoskeletal components), extracellular (caused by the perturbation of extracellular matrix) or as simple as the variation of the tension of the cell membrane. Additionally, it is important to note that some channels, although not gated directly by mechanical perturbation, are modulated by it. Recent studies have been pointing to emerging candidates that are sensitive to biophysical changes of the lipid bilayer that surrounds them [7].

The most well characterised mechanosensitive channels are from the bacterium *E. coli*. In prokaryotes these channels are gated exclusively by membrane deformation and are often used as a paradigm in the study of the effect of mechanical force across the membrane [2, 5, 8]. Since bacteria are easy to manipulate experimentally, these channels have been extensively cloned and their 3D crystal structure has been solved. In the following sections, we use this and other data available to develop a theoretical approach to study how this type of channels can be affected by their spatial surroundings. These channels locally deform the membrane in the process of their gating, and simultaneously their gating is a consequence of the membrane deformation [9]. Therefore the gating of a given channel will be a direct result of the bilayer stretch (due to turgor pressure) or a reaction to other mechanosensitive channels in its neighborhood. It is the final aim of this chapter to understand how these membrane-mediated interaction may lead to a cooperative behaviour, and how this alters the channel gating.

We follow with a more detailed description of the function and properties of these channels in prokaryotes and eukaryotes.

3.2.1 Prokaryotic Mechanosensitive Channels

In prokaryotes the mechanosensitive channels have a vital function in osmotic regulation. It was noticed that bacteria can easily adapt to a wide range of osmolarities and even survive strong osmotic shock [2, 6, 8]. For *E. coli* bacteria, a certain membrane tension is necessary for growth and cell division. The usual pressure felt by the membrane is in the range of 2–6 atm [2]. When *E. coli* is placed in a medium of low osmolarity, water enters the cell. Strong water influx increases the pressure on cell membrane and the cell rapidly increases in volume. The membrane turgor increases, activating the mechanosensitive channels that feel the bilayer stretch. These channels allow for the passage of water, together with its solutes, to quickly equilibrate the osmotic levels of the cell with the environment (see Fig. 3.1).

3.2.1.1 Mechanosensitive Channels of Large Conductance

Large mechanosensitive channels were first isolated from *E. coli* and are the most well understood mechanosensitive channels. The reconstruction of the channel in lyposomes and native membrane patches revealed most of its properties. However the

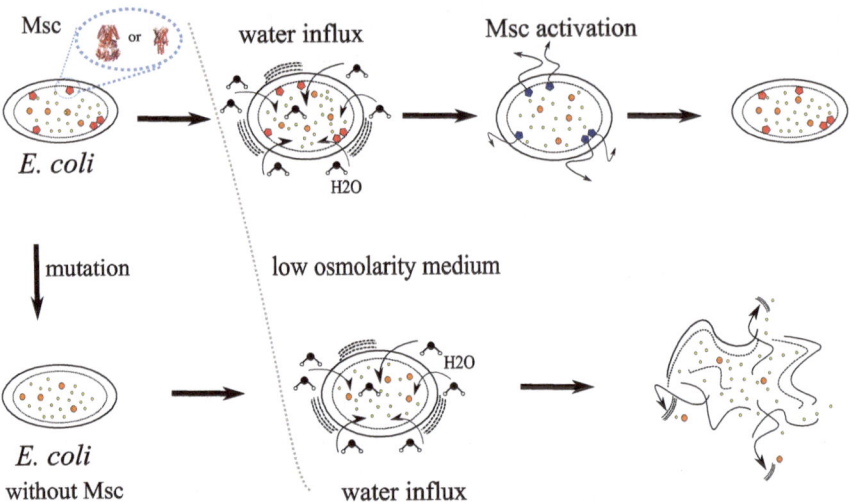

Fig. 3.1 Reaction to osmotic shock of *E.coli*. When immersed into low osmolarity medium the water penetrates into the cell increasing its volume. With the rise of internal pressure the membrane and the cell wall are stretched to the new size. The tensioned membrane exercises a force on the channels that leads to their opening. The open Msc release internal solutes with water from the cell and the pressure is relieved

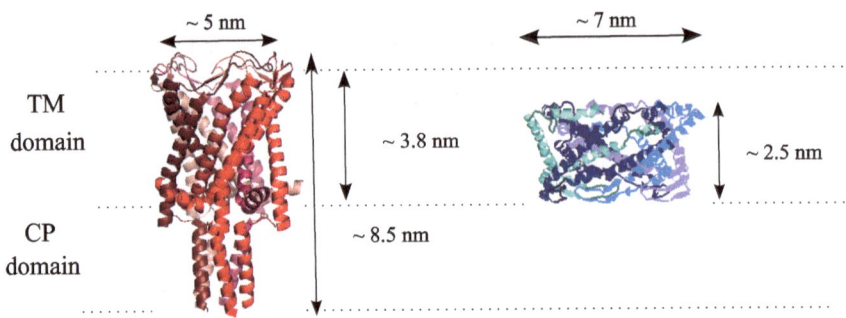

Fig. 3.2 Closed and open structure of MscL (from the crystal structure 2OAR and trans-membrane domain of the open structure simulated by Sukharev et al. [11]) viewed with PyMOL. TM corresponds to the trans-membrane domain of the protein and CP is the cytoplasmic domain

first crystal structure was resolved for MscL of *M. tuberculosis* [10]. In this work the channel was crystallised in its closed state, revealing a homomeric protein complex composed by five subunits, see Fig. 3.2. The channel has most of its volume immersed in the lipid bilayer (the trans-membrane domain) and has only a small part inside the cytoplasm (the cytoplasmic domain). Each of the channel subunits possesses two trans-membrane helices: TM1 forms the pore and TM2 is external and enters in contact with lipids. The gating mechanism of the channels involves a rotation of those helices. Therefore there are not many bonds linking these helices together, and they have some freedom to move without constraints [11]. MscL from *M. tuberculosis* is homologue to MscL of *E. coli* and experiments suggest that they do function by similar mechanisms. Therefore it was possible to use the obtained crystal structure which served as a template to establish a model for the structure of MscL of *E. coli*. Works by Sukharev et al. [11, 12] on molecular dynamics proposed also the open state configuration structure. The diameter of the open pore was estimated in previous experiments and established to be around 4 nm [13]. The channel conductivity was measured to be around 3 nS [14]. Which gives a water flux estimation through the pore to be $\sim 10^8$ molecules/s, which corresponds to 0.1% of the cell volume per second. Furthermore a pore of this size is large enough to be able to transport small proteins (of around 10 kD). The loss of proteins can be lethal for the cell. Therefore it is speculated that MscL is used only as a final resource to avoid cell death in the conditions of osmotic stress. For this reason the channel gating tension of MscL is much higher than other mechanosensitive channels of bacteria. The gating tension is estimated to be $2.5 \, k_b T/\text{nm}^2$ [6, 15], which corresponds to a energy of activation of ~ 50 kJ/mol. To avoid unnecessary leakage, the channel should also be regulated to close as fast as possible, once the desired osmolarity is restored. The number of MscL channels on the membrane was estimated to be ~ 5. In such small concentration they can rarely interact, however in the vast majority of studies they are overexpressed or inserted in to lyposomes in high concentrations.

3.2.1.2 Mechanosensitive Channels of Small Conductance

For reasons still not well understood, there is more than one type of mechanosensitive channel in microorganisms [2]. In prokaryotes, another such channel is the MscS, which it is very different structurally from MscL. A possible common ancestor of both channels was suggested to be a type of channel from archae [6]. The crystal structure of MscS in *E. coli* (see Fig. 3.3) was solved by Bass and coworkers (in its closed state) [16] and by Booth (in the open state) [17]. These works show the small mechanosensitive channel as a homomeric protein composed by five subunits with a pore size of ~ 1 nm. It has three trans-membrane helices and a large cytoplasmic domain. The number of channels present in a prokaryotic cell was estimated to be around 20. Homologues of MscS are present in several other microorganisms, and as well as in all plants [7]. The gating threshold was established be $\sim 1 \, k_b T/\text{nm}^2$ [15].

3.2.2 Eukaryotic Mechanosensitive Channels

3.2.2.1 Mechanosensitive Channels in Animals

In animals there are many types of cells that are found to have some type of mechanosensitive channel analogue. These channels have a different evolutionary origin from MscL and MscS of prokaryotes, however they have very similar way of function [7]. The similarity lies on the influence of the membrane on the channel's gating process, although their complete function normally involves a more complex mechanism. For example, it can involve the cytoskeleton or even the extracellular matrix. These channels can be adapted to very distinct and specific types of functions [7, 8]. In sensory cells they are responsible for identifying mechanical inputs such as sounds (hearing), and gravity (balance). They are also responsible for the process of touch sensation, proprioception. Additionally they are important for body homeostasis, such as in the case of blood pressure, and also morphogenesis and cell migration. Examples of such channels are: TRP channel family, MEC and K_{2P}.

3.2.2.2 Mechanosensitive Channels in Plants

Although there are no proteins that are evolutionarily related to bacterial MS in animals, in plants the situation is different, and there are homologues to MscS channels which are evolutionarily linked to their bacterial counterparts [7]. Although they were established to be sensitive to tension in the membrane, their function is not yet characterised and can be only inferred from the bacterial analog. Plants do respond to several mechanical stimuli and many works are still in progress to relate the activities of MS to those responses [5]. Mechanosensitive channels play an important role in the gravity sensation in plants. The growth in the appropriate direction is possible as a response to a sensory system in the tip of the root and in the endodermis of the stem [5]. Also mechanosensitive channels are important for osmotic regulation. It was also proposed that MS channels can be responsible for the temperature responses in plants, which activate different genes with changes in climate. The mechanism proposed for activation involves properties of the channels that make them sensible to membrane fluidity.

3.3 Experimental Techniques Employed in the Study of Mechanosensitive Channels

The electrophysiological techniques are usually employed together with genetic and biochemical methods in the study of mechanosensitive channels [2]. The channel structure could be obtained as described previously for small and large mechanosensitive prokaryotic channels. The identification of gene of these bacterial channels allowed for their cloning and sequencing. It was also possible to obtain mutations responsible for the gain-of-function and loss-of-function for the channels. These studies permitted to understand better the gating mechanism.

Another subject of study has been the channel interaction with lipids. For this purpose it was possible to use the fluorescence of Trp residues (a fluorophore), which requires a mutation of the channel by the addition of Trp or Cys residues. These studies allowed to determine the preference for lipid of mechanosensitive

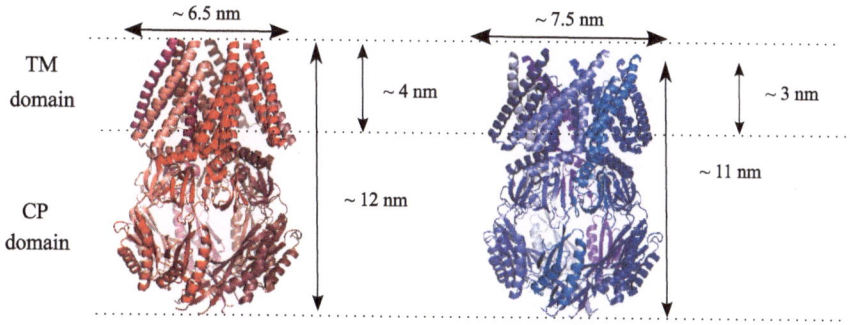

Fig. 3.3 Crystal structures of MscS in closed (*left*) and open (*right*) conformations (PDB codes 2OAU and 2VV5, viewed with PyMOL)

channels. For *E. coli* membrane in particular, it was found no evidence for a strong lipid preference [18].

Another type of study analysed the lipid composition effect on the channels response to osmotic stress [19]. They find strong evidence that changes in the lipid environment does affect the gating. The variation of the membrane environment composed by lipid hydrophobic chain from 16 to 20 carbons increases the gating tension threshold of the channels.

In the following two sections we summarize the important aspects of the two main techniques that we consider to be the core of the research necessary to validate our model. The Sect. 3.3.1 describes the mechanism to measure the tension sensitivity of mechanosensitive channels and the Sect. 3.3.2 allows for channel visualisation on the membrane.

3.3.1 Measurement of the Channel Gating and Sensitivity to Tension

The patch clamp method uses a micro-pipette to record currents trough ion channels. The imposition of a pressure differential across the bilayer drives ion movement, which permits to record the changes between the open and the closed states of the ion channel. This measurement can be used to establish the channel sensitivity to pressure (and tension). Furthermore, examination of the step changes in the ion current provides the channel opening time and conductance. Additionally, this experiment allows to determine the ion selectivity.

However, the bacterial cells are too small to be analysed directly by patch-clamp methods. The diameter of a *E. coli* cell is $\sim 3\,\mu$m when the patch clamp pipette has a diameter of ~ 3–$10\,\mu$m. Therefore the first significant achievement in this field was the development of methods to overcome this difficulty [6]. One technique made possible to fuse the native bacterial membrane with liposomes or even

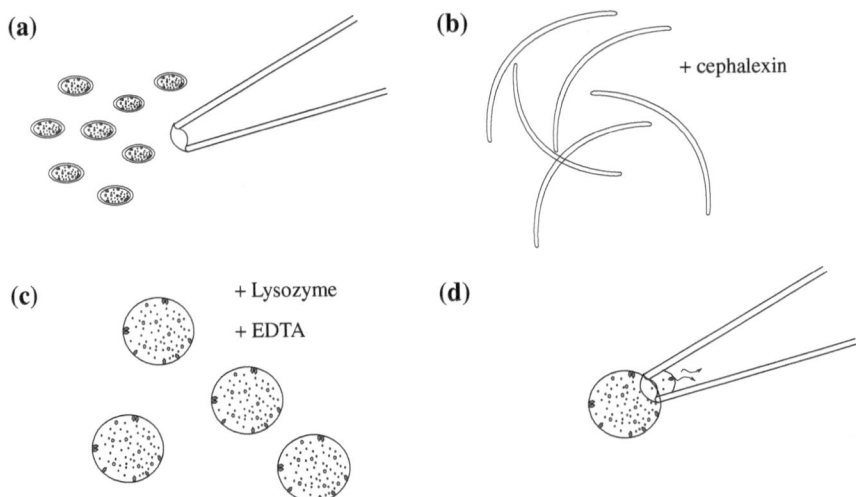

Fig. 3.4 The preparation for the patch-clamp in bacteria. **a** the bacteria *E. coli* is smaller than the patch-clamp pipette. **b** growing the bacteria with the presence of cephalexin, the cells division is inhibited **c** Lysozyme and EDTA digest the cell wall **d** the produced spheroplasts are large enough to allow the patch-clamp measurements

to directly insert the purified channels in liposomes. Another type of essay developed by Martinac et al. [20] 20 years ago made possible to grow giant spheroplasts from *E. coli* cells. The method involved the use of the antibiotic—Chephalexin, that inhibits cell division. In the presence of this substance bacteria cells grow in snake-like filaments and are unable to divide. After the growth their cell wall is digested with lysozyme and EDTA producing spheres with radius $\sim$5 μm, called spheroplasts (see Fig. 3.4). These spheroplasts are then large enough to enable further studies [20].

3.3.2 Spatial Localization of Mechanosensitive Channels

To study functional localization, one needs to know how the protein is distributed on the cell membrane. The usual and most extensively employed technique is to fuse the protein gene with the gene of GFP (green fluorescent protein) and express it. After expression the protein can be visualised by confocal microscopy. This method was used to obtain the fusion protein MscL-GFP in bacteria by Norman et al. [21]. This study shows that channels are formed on the membrane, and compares them with the wild type. Their main properties can be summarized as being very similar to wild MscL, however with a slightly higher gating threshold. The visualisation of MscL-GFP is the done using confocal microscopy. This study shows the spatial distribution of mechanosensitive channels on lyposomes and it is possible to see that their spatial distribution is not homogeneous [21].

This summarizes the most important techniques that are necessary for the study and determination of all the necessary parameters of our study. In the next section we explain some of the results of the localization of mechanosensitive channels on the membrane and describe some evidences for their agglomeration.

3.4 Evidences of Clustering of Mechanosensitive Channels

The interaction of mechanosensitive channels on the native membrane of E. coli could not be observed. These channels are present in low densities (~5 per cell) and the interactions should be very rare events. However the artificial environment created when experimental studies are performed can favor the occurrence of interactions if the number of channels is increased. In fact, in many of studies performed in bacteria those channels are overexpressed. Also in some studies described previously in Sect. 3.3.2 there were observations of non-homogeneous distributions of large mechanosensitive channels in lyposomes [21].

Although there is no direct evidence for cluster formation in native membranes of bacteria, in eukaryotes it was observed for several distinct organisms. For example, there are reports on the clustering of mechanosensitive channels in plants, where special MscS-like proteins called MSL2 and MSL3 cluster on the poles of plastids in the plant A. thaliana [5, 22]. The studies speculate that this distribution favors the redistribution of the channels during the plastid fission. However our study suggests that the function of the channel will also be altered by their group reaction behaviour.

Another reported case of channel aggregation is the clustering of mechanosensitive channels such as MEC-4 and MEC-10 in C. elegans [23]. These channels are responsible for the sense of touch in this organism and form equally spaced domains in neurons [4]. This model of mechano-transduction is still not completely understood and involves a complex of proteins associated to microtubules. However, we speculate that the role of clustering and cooperative gating are also relevant in this case.

3.5 Individual Channel Gating

As described previously, our objective is to study the effect of cooperative behaviour on the function of mechanosensitive channels. However, it is important first to describe the function of an isolated channel.

Mechanosensitive channels switch conformations between an open and a closed state as a response to the tension in the membrane, a process which is denominated channel "gating" [2]. For a single channel, we can write the conformational energy as a linear function of tension,

$$\Delta G = \tau \Delta A - \Delta G_0, \tag{3.1}$$

where ΔG_0 is the energy difference between the closed and open conformations ($50\,k_b T_0$ for MscL), in the absence of an externally applied membrane tension, and ΔA is the difference in area of occupied by an open and a closed state ($20\,\text{nm}^2$ for MscL) [24]. From this expression for the energy, we can represent the probability (P_0) to find a channel in the open state, according to the Boltzmann distribution,

$$\frac{P_0}{1 - P_0} = e^{\frac{\tau \Delta A - \Delta G_0}{k_b T}}, \tag{3.2}$$

where k_b is the Boltzmann constant, and T is the temperature [8]. According to Eq. 3.2, the channel should open with larger probability at the gating tension $\tau_0 = \Delta G_0 \Delta A$($2.5\,k_b T_0/\text{nm}^2$ for MscL), and the transition should be more abrupt for smaller temperature values.

Having described the isolated mechanosensitive channels and their function we turn to the theory of lipid–protein and protein–protein interactions, in the next section. We will see in the following sections that when the most relevant types of interactions are considered, the process of channel gating becomes much more elaborate.

3.6 Interactions Between Membrane Inclusions

Since the development of the mosaic fluid model, the membrane is usually modelled as a two-dimensional fluid consisting of lipids. Membrane inclusions (i.e. bodies inserted in the lipid bilayer that are not lipids themselves, such as proteins or enzymes) diffuse and rotate in this fluid. However, as we already pointed out before, this diffusion dynamics is highly complex, and the diffusion coefficient depends on local diverse compositions, due to the crowded environment on the membrane. In such an environment, the mutual interaction among active membrane constituents plays an important role in the determining of its effective "fluidity", which is an important factor in many physiological processes. Therefore, it is necessary to modify the mosaic fluid picture, by considering not only diffusion but also the interaction amongst inclusions. In this part we give a summarized overview of the types of interactions which are relevant in this context.

The interactions between membrane inclusions could be divided in two types: Direct and indirect. Direct interactions arise due to the presence of an electric charge. The indirect forces, on the other hand are more elaborate, and are mediated by the membrane. They can either be elastic, resulting from the deformation of the membrane, or entropic when they are due to thermal fluctuation of the membrane. A summary of all relevant forces is shown in Table 3.1.

Table 3.1 The forces	Electrostatic	Elastic	Entropic
between membrane inclusions. The electric force is a direct force of short range. The forces mediated by the membrane can be further divided in entropic and elastic	Coulomb Repulsive Short range $\sim \exp(-r)$	Elastic Attractive/repulsive Short range (thickness) long range (midplane) $\sim \exp(-r)$	Casimir-like Repulsive Long range $\sim 1/r^4$

3.6.1 Direct Protein–Protein Interactions

Proteins will directly interact due to their electric field. The electric force between two proteins is usually repulsive considering that proteins are on average equally charged. The electrostatic field of a charge Q at a distance r is given by [25]

$$E(r) = \frac{Q}{4\pi \epsilon_0 r^2}. \tag{3.3}$$

However, in an ionic solution there is an effect known as screening of electric charges. This phenomenon can be described as a rearrangement of free ionic charges in the solution which tends to neutralize the long-range electrostatic field. The theoretical approach used by Debye and Huckel [25] explained the exponential decay of the electric field in the ionic solution, which leads to a shielded electric field,

$$E(r) = \frac{Q}{4\pi \epsilon_0 r^2} e^{-\sqrt{\frac{\rho e^2}{\epsilon_0 k_b T}} r}, \tag{3.4}$$

where k_b is the Boltzman constant, T is the temperature, ϵ_0 is the dielectric constant, e is the unit charge and ρ is the density of charges in solution. Thus, electric forces decay exponentially with the distance, and the range in which we could consider the electric interaction as significant is given by the Debye length, which corresponds to

$$\lambda_D = \frac{1}{k_0}, \tag{3.5}$$

where

$$k_0 = \sqrt{\frac{\rho e^2}{\epsilon_0 k_b T}}. \tag{3.6}$$

For a physiologic solution (0.1 M), we have $\lambda_D = 1$ nm. However it is considered that for small ions the electroneutrality holds for even smaller distances [26]. Therefore, these forces are always short range, and can be considered as extremely weak for proteins on the membrane.

3.6.2 Membrane-Mediated Protein–Protein Interactions

Some of the interactions arise and propagate trough the membrane and depend strictly on the bilayer properties. We can divide such interactions in two main types accordingly to their origin: (1) entropic forces, which arise due to the thermal fluctuation on the membrane, and leads to long range attractive forces; and (2) elastic forces, which result from the elastic deformation of the lipids.

We follow by reviewing the entropic and elastic forces in detail. We focus on elastic forces since they are the most relevant in the study of mechanosensitive channels. We then describe the elastic theory approach used to determine the energy of interacting channels in closed and open states.

3.6.2.1 Entropic Interactions

The entropic interactions appear as a consequence of membrane thermal fluctuations, which occur in physiological temperatures. The spectrum of vibration of the membrane depends on the distance between the proteins, and the free area between them. To increase the entropy, the system tries to maximize the number of modes of those fluctuations. This effect is the classical analog to the Casimir effect [27] in quantum electrodynamics, which predicts an attraction between conducting plates mediated by (quantum) fluctuations in the electromagnetic field. The resulting effect of these fluctuations is an attractive potential between two objects at distance r, which decay as $1/r^4$ [28]. This constitutes a long-range interaction, which is proportional to $k_b T$ and scales with the square of the area A of the inclusions [29] as,

$$V(r) \sim -\frac{A^2}{r^4} k_b T. \tag{3.7}$$

It is important to emphasise that this force will not exist at short ranges because of its stochastic origin. The membrane is not a continuous matter but is composed by discrete lipids, and it is not possible to have such fluctuations distances of only few lipids between proteins.[1] For this reason this type of force is not relevant in crowded biological membranes, such as in the wild type *E. coli*. However this interaction can be important in uncrowded membranes such as in liposomes. Furthermore, the amplitude of this interaction is not very large. For instance, considering a mechanosensitive channel of large conductance, which has an area of $\sim 20\,\text{nm}^2$, for distances larger than 4 nm (~ 4 lipids) the interaction is already smaller than $1\,k_b T$.

3.6.2.2 Elastic Interactions

When a protein is inserted in a lipid bilayer it disturbs the bilayer away from its unperturbed, relaxed state. The usual approach to treat this type of system is to

---

[1]   The diameter of a single lipid is approximately 0.75 nm.

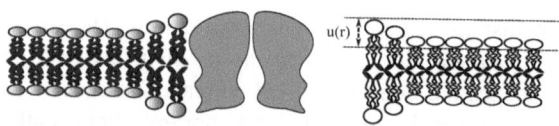

Fig. 3.5 Schematic representation of a thickness deformation. The important parameter is $u(r)$, that characterizes this deformation profile around the inclusion

consider the protein as a rigid inclusion and the surrounding membrane as an elastic medium that deforms to accommodate it. The idea behind this is that the bilayer matches its hydrophobic core with the hydrophobic part of the protein. The result is an exponentially decaying perturbation around the inclusion.

There are two analytical approaches used to describe elastic forces on the membrane: The Landau-de-Gennes theory and the Elastic theory [30]. Both methods describe the system (bilayer and inclusion) through its free energy. The first method is derived from works on liquid crystals and consists of a Landau form for the expansion of the energy functional in the order parameter [31]. In this case the order parameter is $u(r)$, which is the difference between the size of the protein hydrophobic core and the size of the lipid chain (see Fig. 3.5). The second formulation is based on a continuum approach of elastic theory that describes the deformation of the lipid monolayers. In this method the inclusion is an additional constraint added through the boundary conditions of the differential equation.

It is also common to use molecular dynamics simulation to describe the protein-lipid interactions and deduce the energy involved in the accommodation of the integral membrane proteins [32–34]. These simulations are performed on several scales, using both fine-grained and coarse-grained approaches. They are very important complement to the analytical studies because of the difficulties that still underlay the experimental tests of the theory.

Here we will use the well-established mathematical theory that describes this system based on the determination of the energetic cost related to the membrane deformation. Although some works in the area include different additional terms for the energy of membrane deformation [30], there are three main properties of the membrane that give it the ability of storing energy upon an elastic deformation [6, 24].

1. *The Membrane has a specific equilibrium hydrophobic thickness.* This yields a resistance to hydrophobic mismatch of embedded proteins. The membrane stretches or compresses to match the hidrophobic core of the protein to the hidrophobic core of the lipids.
2. *Bending stiffness of the membrane.* The membrane tends to resist to changes in angle between adjacent molecules. In other words, the membrane has a resistance to curvature.
3. *Preferred spacing between lipid molecules.* The membrane is highly resistant to volume changes. Therefore, if the protein extends the membrane vertically, it will compress laterally. This resistance is proportional to the tension of the membrane.

The protein inclusions can have very diverse shapes. The shape of the protein is introduced as a boundary condition in the problem above [6]. The two most commonly used shapes are conical and cylindrical. The conical shape bends the membrane, due to its vertical asymmetry, and imposes what is called the midplane deformation. On the other hand, proteins with cylindrical shape impose only a thickness deformation. For mechanosensitive channels in bacteria, the trans-membrane domain of MscL will account for a thickness deformation, and of MscS will account for both thickness and midplane deformations.

In the next part we derive the calculation of the two types of elastic deformation (thickness and midplane), and apply them to mechanosensitive channels of large conductance of *E. coli*.

3.6.2.3 Interaction Energy Between Mechanosensitive Channels

The elastic theory approach is formulated as a variational principle for the free energy functional $G(u(r))$ [30]. The first step is to formulate the terms which are included in this functional. There are numerous works on the subject, which include different possible degrees of freedom for the deformation of the membrane, and account for different boundary conditions. Here we present the approach of Ursell et al. in [24], which considers the hydrophobic mismatch (first term of Eq. 3.8), the membrane bending (second term of Eq. 3.8) and the conservation of the volume of lipid molecules (last term of Eq. 3.8):

$$G = \int \left[\frac{K_a}{2} \left(\frac{u(r)}{d_0} \right)^2 + \frac{\kappa_b}{4} (\nabla^2 u(r) - c_0)^2 + \tau \frac{u(r)}{d_0} \right] dr^2, \qquad (3.8)$$

where the integration is taken on the two-dimensional membrane surface, and the constants are the stretch modulus K_a, the bending modulus κ_b, the spontaneous curvature c_0, the tension τ and the lipid length d_0. Here we use a trick to transform the equation to a more convenient form by summing τ^2 times membrane area, which is identically zero when calculated for differences in the free energy.

$$G = \int \frac{1}{2} \left(K_a \left(\frac{u}{d_0} + \frac{\tau}{K_a} \right)^2 + \kappa_b \left(\left(\frac{\partial^2 u}{\partial x^2} \right) + \left(\frac{\partial^2 u}{\partial y^2} \right)^2 \right)^2 \right) d\Omega. \qquad (3.9)$$

For $G(u)$ represented as

$$G = \int L \left(u, x, y, \frac{\partial^2 u}{\partial x^2}, \frac{\partial^2 u}{\partial y^2} \right) \qquad (3.10)$$

we can proceed with the variational approach [35] to find $u(r)$ that minimises $G(u)$.

$$\frac{\partial L}{\partial u} = \frac{2K_a}{d_0^2} \left(u + \frac{\tau d_0}{K_a} \right) \qquad (3.11)$$

$$\frac{\partial L}{\partial(\partial^2 u/\partial x^2)} = 2\kappa_b \left(\frac{\partial^2 u}{\partial x^2} + \frac{\partial^2 u}{\partial y^2}\right) \quad \frac{\partial}{\partial^2 x}\frac{\partial L}{\partial(\partial u/\partial^2 x)} = 2\kappa_b \left(\frac{\partial^4 u}{\partial x^4} + \frac{\partial^4 u}{\partial x^2 \partial y^2}\right)$$

(3.12)

$$\frac{\partial L}{\partial(\partial^2 u/\partial y^2)} = 2\kappa_b \left(\frac{\partial^2 u}{\partial x^2} + \frac{\partial^2 u}{\partial y^2}\right) \quad \frac{\partial}{\partial^2 y}\frac{\partial L}{\partial(\partial u/\partial^2 y)} = 2\kappa_b \left(\frac{\partial^4 u}{\partial y^4} + \frac{\partial^4 u}{\partial x^2 \partial y^2}\right)$$

(3.13)

This results in a differential equation

$$\nabla^4 u + \frac{K_a}{\kappa_b d_0^2} u + \frac{\tau}{K_a \kappa_b d_0} = 0.$$

(3.14)

The solution of this differential equation with the necessary boundary conditions gives the deformation profile of the membrane. The boundary conditions are determined by the dimensions of the inclusion, i.e. by the hydrophobic mismatch and the slope at the protein interface. Consider r_i the coordinates of the border of the inclusion. It is necessary to specify the mismatch $u(r_i)$ between the lipid core and the core of the protein bilayer and the slope $\nabla u(r_i)$ of the bilayer at the protein–lipid interface. The slope will be zero for thickness deformation ($\nabla u(r_i) = 0$), and will correspond to the deformation angle for a midplane deformation ($\nabla u(r_i) = \theta$).

After the determining the deformation profile $u'(r)$ it is necessary to obtain also the energy of the deformation, which is given by Eq. 3.8. The energy cost of the inclusion is therefore

$$E(r) = G(u'(r)).$$

(3.15)

In the next part we solve numerically the equations derived with parameters of MscL to calculate the interaction force between the channels in different states.

3.6.2.4 Interaction Between MscL Channels

The MscL channel from E. coli deforms the membrane with a thickness deformation (Fig. 3.6). Using the Eq. 3.14 and the parameters presented in Table 3.2, we can calculate numerically the deformation of the membrane around two channels, which are placed at distance r between each other (Fig. 3.7).

We have to consider the channels in the system in two possible states, open or closed, since the parameters for both states are different (see Table 3.3). In Fig. 3.12, we can see the resulting interaction energies for two channels, as a function of the distance between the channels. It shows that two channels in the same states (open–open, closed–closed) attract each other, although with different intensities, with the attraction between open channels being the strongest. On the other hand, the channels in different states (open–closed) always repel each other.

Because of the non-linearity of elastic forces, it is not possible to infer the many-body elastic interactions directly from pair-wise forces. However, it appears that

Table 3.2 Membrane properties for *E. coli*. The parameters are the stretch modulus K_a, the bending modulus κ_b, the tension τ, the lipid length d_0 and the temperature T

Parameters	
K_a	$58\,k_bT/\text{nm}^2$
κ_b	$14\,k_bT$
τ	$2.6\,k_bT/\text{nm}^2$
d_0	$1.75\,\text{nm}$
T	$300\,\text{K}$

Table 3.3 Boundary conditions for the large mechanosensitive channel

State	r_0 (nm)	u_0 (nm)
Closed	2.5	0.15
Open	3.5	−0.5

the addition of individual pair-wise interactions is sufficient to describe the many-body system with good precision. This can be seen in Fig. 3.8, where the result for a system of three channels is compared with the composed energies of a system of two channels. Since the differences are very small, this justifies our following assumption that the many body interaction can be considered as juxtaposition of individual pairwise potentials.

Using the knowledge obtained about the pair-wise interactions of mechanosensitive channels, in the following section we formulate a coarse-grained model for its collective behaviour, from which we can derive the consequences of cooperative behaviour.

3.7 Model of the Cooperative Gating of the Mechanosensitive Channels

As we have discussed in the previous sections, the most relevant forces governing the behaviour of mechanosensitive channels are short-distance elastic interaction mediated by the membrane. Using this information, in this section we develop a model which describes the channel aggregation and the collective gating of aggregated channels. Since the interactions are short range, and we are interested in the collective behaviour of a large number of channels, we chose a discretized approach for the development of the model, where individual channels are placed on a two-dimensional square lattice, and interact only with the four nearest neighbors. Additionally, we divide the model in two parts (see Fig. 3.9): first, we describe the spatial distribution of these channels on the cell membrane in normal physiological conditions, see Sect. 3.8. At this point all the channels are found only in the closed state and are allowed to move freely on the lattice, interacting by short range attractive forces, as determined previously in Sect. 3.2.6.4. Secondly, after obtaining the equilibrium spatial configuration that these channels assume on the membrane, we depart from the physiological conditions, and consider the channel gating, Sect. 3.9. Since the gating occurs in a much shorter time scale than the diffusion, we consider that the channels are no longer allowed to move at this point. Each channel is allowed

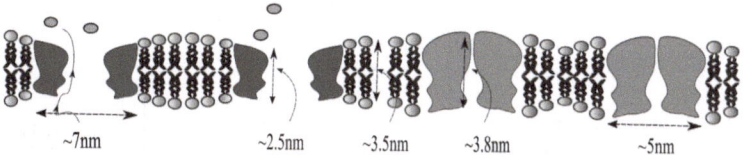

Fig. 3.6 The mechanosensitive channels of large conductance stretch the lipids around when in the closed state. In the open state however the channels compress the membrane bilayer

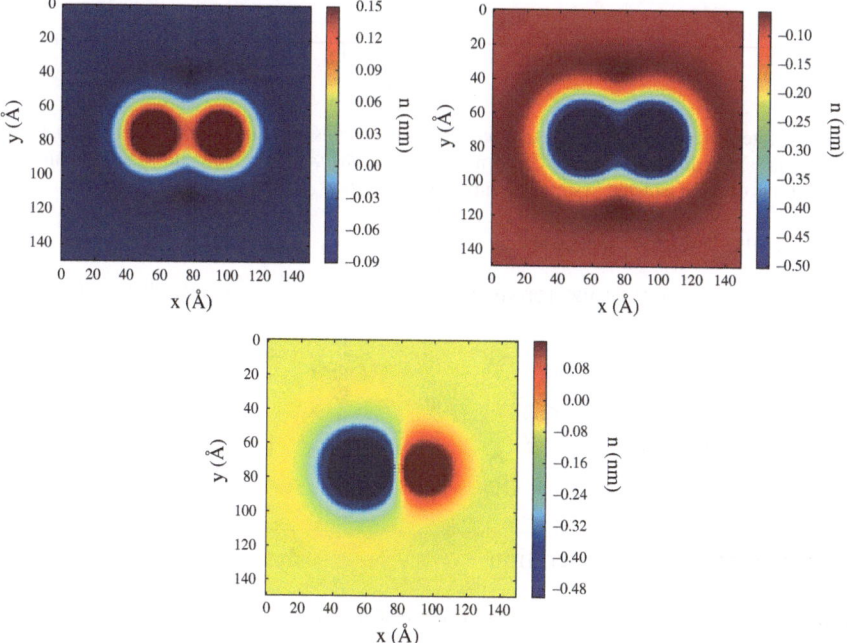

Fig. 3.7 The deformation profile u around two open channels (*top left*), two closed channels (*top right*) and an open and a closed channel (*bottom*)

to be in two possible configurations: open or closed, and their interaction energies correspond to what was calculated in Sect. 3.6.2.4.

3.8 Dynamics of Agglomeration

In this part we analyse the conditions for non-homogeneous distribution of mechanosensitive channels in a single conformation. The channels are distributed on a lattice, with periodic boundary conditions, which represents the cell surface. At this stage we consider the conditions of the cell previous to the osmotic shock in the optimum conditions of growth. In these conditions the membrane is subjected to a

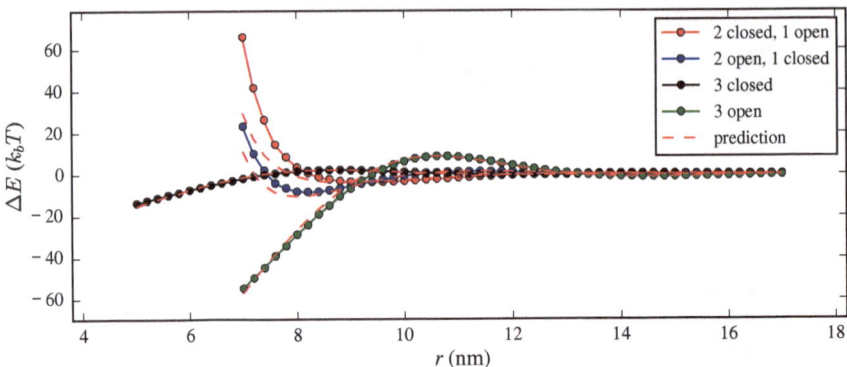

Fig. 3.8 Interaction energy between three equidistant channels, as a function of the distance between their centers, for different channel conformations. The *dashed lines* show the summed pair-wise interactions of three isolated systems of only two channels

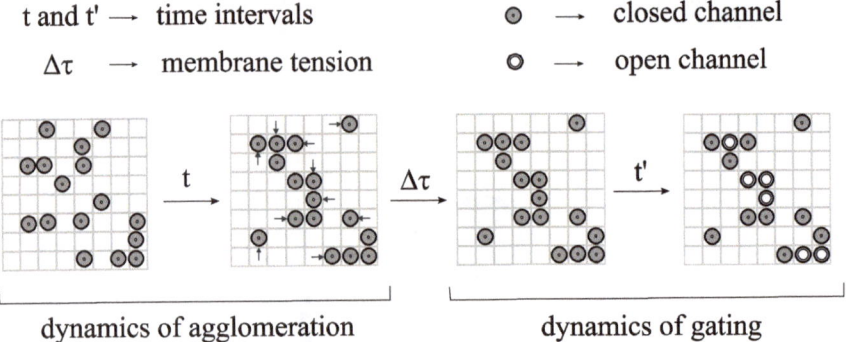

Fig. 3.9 The model is divided in two parts: (1) the dynamics of agglomeration; and (2) the dynamics of gating

very small net tension of $0.5\,k_b T$ [2] and all the channels are spread on the membrane in the closed state. We assume that channels sitting on neighbouring sites interact trough elastic forces. The interaction of neighbouring forces results in a decrease in the global energy of the system. This can be described by the following Hamiltonian

$$H = -J \sum_{\langle ij \rangle} s_i s_j, \tag{3.16}$$

where the sum is taken over neighboring sites, and J is the interaction strength deduced from elastic interactions in the previous section, which is $\sim 1.25\,k_b T$. The state s_i represents the state of a lattice site (is -1 for an empty site or $+1$ for an occupied site).

This system corresponds to the well known lattice gas model [36, 37]. In the next Sect. 3.8.1 we describe the model and its main properties. We follow with the consequences of this theory for the agglomeration of mechanosensitive channels on the bacteria membrane.

3.8.1 The Lattice Gas Phase Diagram

The lattice gas is an Ising model with a conserved order parameter, given by the density, ρ. A good introduction to the theory is given in [36]. The Hamiltonian of the lattice gas is $H_{\text{lattice}} = -\epsilon \sum_{\langle ij \rangle} \theta_i \theta_j$. In this expression the parameter θ assumes values $0, 1$ for an empty or occupied lattice respectively, and the energy varies between 0 or $-\epsilon$ for pairs of empty-occupied (empty-empty) or occupied-occupied lattices. This is a simple form to represent a short range attraction, which is exhibited by a decrease in the system's energy for the configurations with larger numbers of aggregated particles. The conserved parameter ρ defines a fixed number of particles on the grid of size N,

$$\sum_i \theta_i = \rho N. \tag{3.17}$$

Under a simple variable transformation $s_i = 2\theta_i - 1$ the H_{lattice} becomes an equivalent of a Hamiltonian of the Ising model, where s_i assumes values $+1$ for occupied and -1 for an empty lattice. Under this transformation the conserved order parameter can be rewritten as

$$M = \sum_i s_i = N(2\rho - 1), \tag{3.18}$$

where the total magnetisation M is fixed, which decreases the number of configurations that can be assumed by the spins, s_i. From this expression we can write the density ρ in terms of the magnetisation per spin $m = M/N$,

$$\rho = \frac{1}{2}(1 + m). \tag{3.19}$$

From the theory of the Ising model and from the Onsager solution [38] we know that m can assume two values below the critical temperature T_c (Curie temperature). Above this temperature, however, the system is disordered and homogeneous. The Curie temperature is, therefore, the point where the Ising model undergoes a phase transition and the system acquires a preferential orientation (spontaneous magnetization). Together with the existence of the above constraint, the two possible magnetisation values compel the system to have also two preferred local densities, which leads to the existence of either a non-homogeneous or a homogeneous phase, depending on the interaction strength J (or the thermal fluctuation T), and on the particle

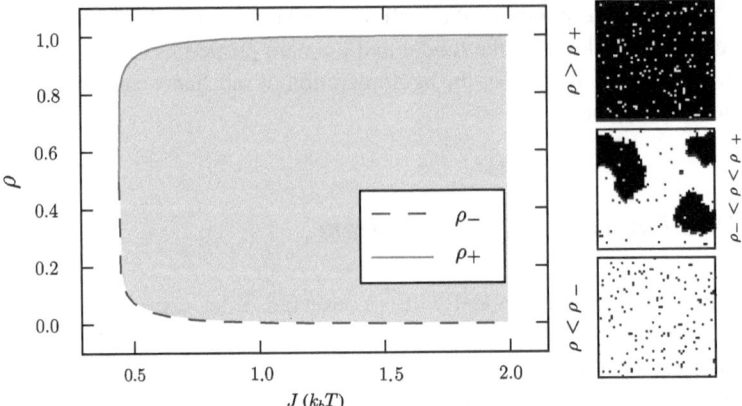

Fig. 3.10 Phase diagram of the lattice-gas model, as a function of the density ρ and the interaction strength J. The system has two possible phases: homogeneous (*white* area) and non-homogeneous (*grey* area). Snapshots of simulations of each phase are shown on the *right*

density ρ (see Fig. 3.10). The system may segregate in two regions, each one with a characteristic density [36]

$$\rho_\pm = \frac{1}{2}\left(1 \pm (1 - \text{csch}^2(2\beta J))^{\frac{1}{8}}\right). \tag{3.20}$$

This equation is a simple combination of the Eq. 3.19 and the Onsager solution [38]. When the density ρ is lower than ρ_- or higher than ρ_+ the particle distribution is homogeneous. However, if $\rho_- < \rho < \rho_+$ the particles segregate in different domains with different local densities: a low density (ρ_-) region and a high density (ρ_+) region (see Fig. 3.10).

The density ρ also gives the area of each region,

$$\rho_+ A_+ + \rho_- A_- = \rho A \tag{3.21}$$

where, A_+ is the area occupied by the high density (ρ_+) region, and A_- is the area occupied by the low density (ρ_-) region and A is the total lattice area

$$A_+ + A_- = A_{\text{total}}. \tag{3.22}$$

Depending on J and ρ the clusters in the high density region can be either compact (for large J) or can be ramified (small J). In Fig. 3.11 we show how the clusters change with the interaction strength J, by measuring the average number $\langle k \rangle$ of occupied sites around a particle in the system. For $J = 0.75\,k_b T$ the aggregates formed are strongly ramified and $\langle k \rangle$ changes strongly with ρ (from 0 to 4). However in the case of $J = 1.25\,k_b T$ the clusters are more compact and $\langle k \rangle$ is already large for small ρ and changes from 3 to 4. In the following sections, where we analyse the influence of the spatial distribution on the gating reaction of the channels, we will show that the compact and ramified distributions have distinct properties.

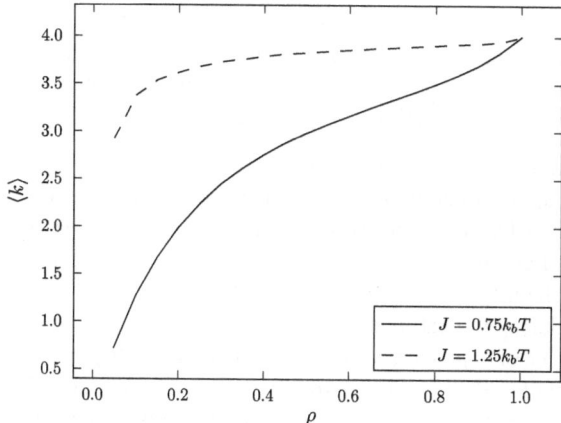

Fig. 3.11 The average number of occupied neighboring sites $\langle k \rangle$ as a function of the particle density ρ, for the lattice-gas model

3.8.2 Conditions for Mechanosensitive Channel Agglomeration

From Eq. 3.20, and using that $J \approx 1.25\,k_b T_0$ (from Eq. 3.8), we obtain that $\rho_- = 1.7 \times 10^{-3}$ channels/site. Channel aggregation will only happen if $\rho > \rho_-$. For a wild type *E. coli* cell there are, on average, 5 MscL channels [2]. The membrane area of this bacteria corresponds to $\sim 6 \times 10^{-12}\ m^2$ [39]. Therefore the density of the channels is $\rho \approx 1.6 \times 10^{-5}$ channels/site, which corresponds to the homogeneous phase, without clustering.

However, there are two ways to induce the aggregation of mechanosensitive channels in *E. coli*. The first one is by changing the attraction between channels, by introducing them in a different lipid environment. With the appropriate value for the parameter J, the non-homogeneous phase could be realised (see Fig. 3.10). The other alternative is by a simple increase of the density of the channels. Thus, the two parameters can be regulated independently.

We will see in the next sections that the gating of the channels depends highly on their distribution on the lattice. In particular, we will see that the more compact are the channels clusters formed, the larger will be the deviation of the collective gating from the individual behaviour.

3.9 Dynamics of Gating

We turn now to the gating response of the channels, when the tension is changed abruptly. This is the case, for example, if the osmolarity of the medium is suddenly decreased. The most important consideration of our approach in this part is that the gating dynamics occurs in a different time scale than the diffusion rate. Therefore in this part of the model, we consider the positions of the channels fixed in the lattice, which correspond to an equilibrium spatial configuration obtained with the lattice gas model of the previous section. We justify this assumption as follows

1. *The gating time is of the order of microseconds.* The gating response of channels is of ~3 μs, approximately microseconds [40]. On the other hand the free diffusion is of the order of ~5 $nm^2/\mu s$, and should be at least 10 times even smaller for a crowded environment, such as a biological membrane [41]. Since the area of a single channel is approximately $\pi(2.5)^2 \sim 20\,nm^2$, they can not diffuse sufficiently during the gating event.

2. *The channel cluster formation itself prevents channels from moving.* Since the interaction of closed channels is attractive, the diffusion rate will be even smaller if the channels are close together. Even those channels on the border of the clusters will have a decreased diffusion escape rate due to this attraction. In the parameter region where the channels do not form clusters, this issue is not relevant since they act independently.

3. *We are interested in the first gating event.* Regardless of the diffusion rate, the assumption that the channels do not move significantly during the gating should be universally valid in our results, since we are mostly interested in the time before the first collective gating event. Until this point, the channels are all closed and governed only by attractive forces. The moment most channels open, the membrane tension changes very quickly, and there is no time for the channels to form an equilibrium spatial conformation with channels in mixed open and closed states.

In this section we are concerned with the following question: how does the spatial clustering, as obtained by the lattice gas approximation, affects the gating response to osmotic tension?

In the following, the channels can assume to possible states: open or closed. We describe the state of each channel i by a variable σ_i, which can have the values $+1$ and -1, corresponding to an open or a closed state, respectively. The state of the channel will depend on the tension of the membrane and also on the state of its neighbors. The new energy of the system can now be written as the sum of the non-interaction energy of the individual channels and their interaction energy.

$$H = H_{non} + H_{int} \tag{3.23}$$

These energies can be obtained by solving Eq. 3.14, for a system of only one or two channels, respectively, and considering all the different channel conformation states. The non-interacting energy from the Eq. 3.23 corresponds simply to

$$H_{non} = h \sum_i \sigma_i, \tag{3.24}$$

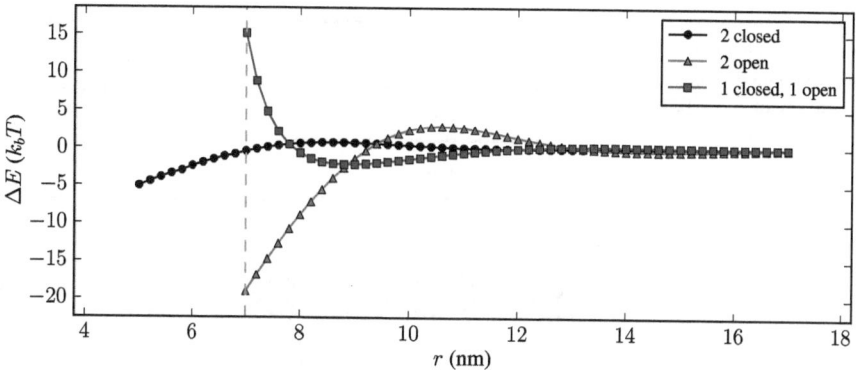

Fig. 3.12 Interaction energy between two channels, as a function of the distance between their centers, for different channel conformations. There is an attractive force between channels when they are in the same state and a repulsive force otherwise [24]

where h is a global non-interaction field

$$h = (\Delta G_{\text{gate}} - \tau \Delta A)/2, \tag{3.25}$$

already described in the Sect. 3.5. The solution of Eq. 3.14 for a non-interacting channel in the open and in the closed configurations results in the mechanical energy necessary for the channel to gate correspondent to $\Delta G_{\text{gate}} \sim 50 \, k_b T_0$. This is equivalent to the value obtained experimentally [24]. The ΔA is the deformation area of the protein, and τ is the membrane tension, which changes according to the osmolarity of the medium.

In an analogous fashion, we can obtain the energy levels for interacting channels. The results were already presented in the Sect. 3.6.2.4, see the Fig. 3.12, where two channels in a similar state have a negative energy contribution and channels in distinct states increase the energy of the system. To calculate the discrete energy levels for our model we assume the following:

1. Since the channels are fixed, we simplify the energy profile of Fig. 3.12 by considering only discrete energy levels corresponding to the interaction at a distance of 6–7 nm between channels centers.
2. Although the interaction energy also depends on the tension, the dependency is not very strong [24], and can therefore be neglected.

The energy levels obtained in this way are shown in Fig. 3.13. These energy levels can be represented in a Hamiltonian form as follows

$$H_{\text{int}} = -\frac{E}{2} \sum_{\langle ij \rangle} \sigma_i \sigma_j - \frac{P}{4} \sum_{\langle ij \rangle} (\sigma_i + \sigma_j), \tag{3.26}$$

considering $E = 10 k_b T_0$ and $P = 10 k_b T_0$, which can be further rewritten as

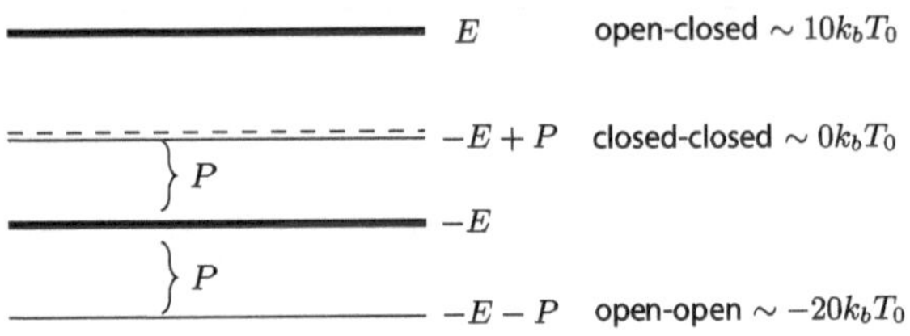

Fig. 3.13 Energy levels for interacting channels, for an approximate distance around 6–7 nm with $E = 10k_b T_0$ and $P = 10k_b T_0$

E open-closed $\sim 10k_b T_0$

$-E+P$ closed-closed $\sim 0k_b T_0$

$\left. \right\} P$

$-E$

$\left. \right\} P$

$-E - P$ open-open $\sim -20k_b T_0$

$$H = h \sum_i \sigma_i - \frac{P}{2} \sum_i k_i \sigma_i - \frac{E}{2} \sum_{\langle ij \rangle} \sigma_i \sigma_j, \qquad (3.27)$$

where k_i is a local field, which is equal to how many occupied neighbors a single channel has, and is thus given by the spatial distribution. As will be seen below the system will behave differently depending on the density of the clusters formed.

In the next sections we explain the properties of this system in equilibrium and the dynamics of escape from an initial metastable state, which we obtain from numerical simulations. For that we employ the Metropolis-Hastings [42] algorithm, which at each step chooses a random occupied lattice site i and flips its state from σ_i to $-\sigma_i$ with a probability given by

$$W(\sigma_i \to -\sigma_i) = \min[1, e^{-\Delta E_i}] \qquad (3.28)$$

with

$$\Delta E_i = \sigma_i \left(Pk_i - 2h + E \sum_{\langle ij \rangle} \sigma_j \right). \qquad (3.29)$$

For the simulations we used a lattice of dimension 1000×1000, a total number of 10^3–10^5 Monte-Carlo steps per site, and each measurement was obtained as an average of 10 independent realizations.

3.9.1 Equilibrium Properties

To study the system in equilibrium, we have to start with the system in random initial conditions, and evolve the system long enough until its macroscopic properties have stabilized. The aim is to understand how the density on the membrane can affect the gating of mechanosensitive channels. Because of the presence of a spatially correlated field k_i the spatial distribution will determine the peculiar properties of this system.

We start with two simplest cases: $\rho < \rho_-$ and $\rho = 1$. In the first case all the channels are non interacting on the lattice. In this situation the channels gate at the tension $2.5 \, k_b T / \text{nm}^2$, as discussed in Sect. 3.5. The next trivial case is the lattice full

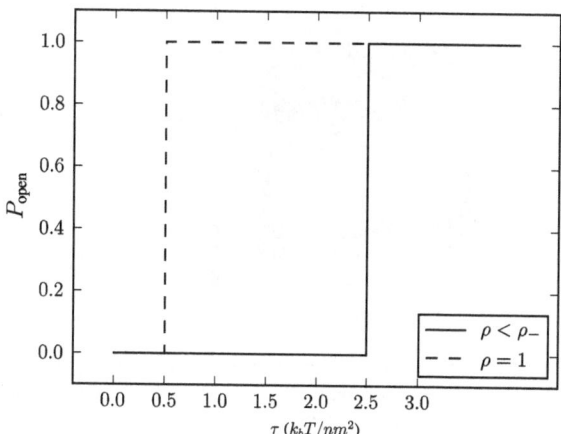

Fig. 3.14 Probability of channel gating as a function of tension, for $\rho = 1$ and $\rho < \rho_-$ (isolated channels). For a system with isolated channels the transition occurs for a tension of 2.5 $k_b T/\text{nm}^2$. For $\rho = 1$, each channel has all four neighbors occupied, and the transition happens for a tension of 0.5 $k_b T/\text{nm}^2$

of channels, each with four occupied neighbors. In this case the system transforms in to a simple Ising model with an external field given by $h + 2P$. In this case the channels gate in a tension of 0.5 $k_b T/\text{nm}^2$, which is significantly lower than the gating threshold of independent channels (Fig. 3.14).

Having examined the trivial cases we proceed our analysis to the intermediate situation where $\rho_- < \rho < \rho_+$. In this case the process of diffusion described in the first model makes the channels agglomerate in clusters of finite sizes with highly irregular structures. Those structures introduce an anisotropy in the field element. This anisotropy makes possible certain configurations where states -1 and $+1$ can coexist in particular fractions. The ratio of channels in a particular state will not only depend on the external field but also on the spatial distribution of channels on the grid. The spatial distribution of the channels on the grid, as was seen in the previous section, depends on the interaction strength J (or the temperature T). Inside the non-homogeneous phase, the cluster compactness will vary gradually with J or ρ, becoming either more ramified (for lower J or ρ) or more compact (for higher J or ρ). In our analysis we will consider as examples $J = 1.25$ for a compact cluster distribution and for $J = 0.75$ for ramified clusters.

In Figs. 3.15 and 3.20 it can be seen the average channel conformation as a function of the membrane tension both for the ramified ($J = 0.75$) and compact ($J = 1.25$) configurations. We can observe several transition steps, from the situation with all channels completely closed, to all channels completely open. Two easily identifiable transitions correspond to the high-density regions (sites with $k_i = 4$), and the low-density regions (sites with $k_i = 0$), for lower and higher tensions respectively. It is also possible to notice some intermediate steps, which correspond to what we could call the "border" of the clusters, i.e. sites with $4 > k_i > 0$. This means that channels with larger number of neighbors tend to equilibrate in the open state, even for small values of tension. As tension is increased, the channels gradually open, from the bulk of the clusters to the "borders".

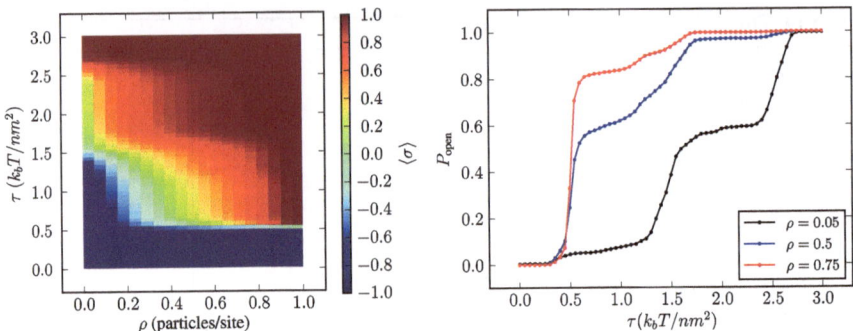

Fig. 3.15 The average channel conformation $\langle\sigma\rangle$ in equilibrium as a function of τ. On the *right* is shown the gating probability P_{open} as a function of the tension τ. The particle spatial distribution is given by the lattice gas with an interaction strength $J = 0.75\,k_b T$

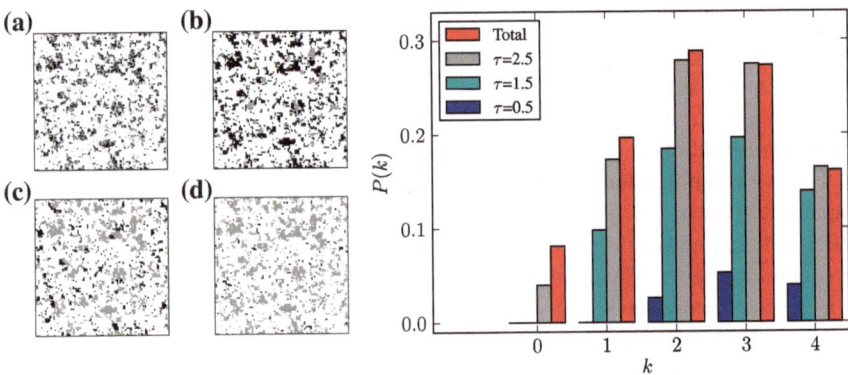

Fig. 3.16 Spatial distribution of open and closed channels for $J = 0.75\,k_b T$, $\rho = 0.25$. On the *right* is the histogram of channels with a given number of occupied neighbors, for both open and closed channels (*red*) and only open channels, for different values of τ. On the *left* are shown snapshots of simulations of the system in equilibrium for a range of tensions τ, with closed channels in *black* and open in *grey*. **a** The initial state of the system before the tension is applied, (corresponds to the red distribution in the histogram); **b** System in equilibrium for $\tau = 0.5\,k_b T/nm^2$ (*grey* distribution in histogram); **c** $\tau = 1.5\,k_b T/nm^2$ (*cyan* distribution in histogram); **d** $\tau = 2.5\,k_b T/nm^2$ (*blue* distribution in histogram)

Now we examine with more detail the case of ramified clusters ($J = 0.75$) and small density $\rho = 0.25$. In this case we have channels with low number of neighbors in average (see the distribution in red in the Fig. 3.16). In this figure we also show the spatial distribution of the open channels in equilibrium, for a series of tension values. It shows how, with an increase of tension, the configuration of open channels move from the "bulk" (large k_i) to the "border" (low k_i). In the case of a larger density ($\rho = 0.5$), the average cluster size is larger (see Fig. 3.17), which means there are more sites in the bulk. Due to this fact, the transition to the open state involves a smaller number of steps as the tension increases (Figs. 3.18 and 3.19).

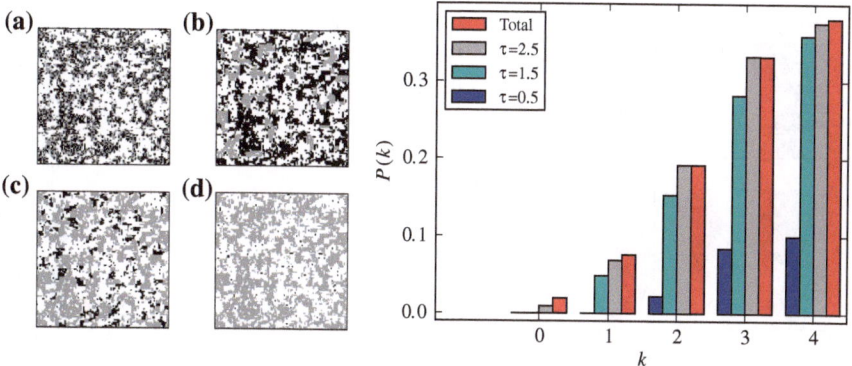

Fig. 3.17 Spatial distribution of open and closed channels for $J = 0.75\,k_b T$, $\rho = 0.5$. On the *right* is the histogram of channels with a given number of occupied neighbors, for both open and closed channels (*red*) and only open channels, for different values of τ. On the left are shown snapshots of simulations of the system in equilibrium for a range of tensions τ, with closed channels in *black* and open in *grey*. **a** The initial state of the system before the tension is applied, (corresponds to the *red* distribution in the histogram); **b** System in equilibrium for $\tau = 0.5\,k_b T/\text{nm}^2$ (*grey* distribution in histogram); **c** $\tau = 1.5\,k_b T/\text{nm}^2$ (*cyan* distribution in histogram); **d** $\tau = 2.5\,k_b T/\text{nm}^2$ (*blue* distribution in histogram)

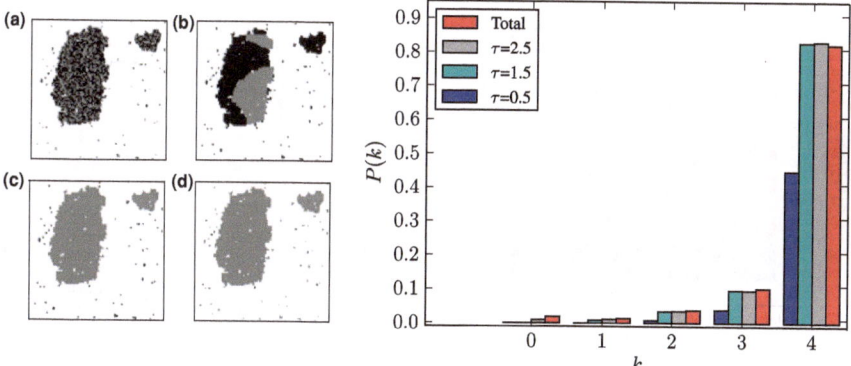

Fig. 3.18 Spatial distribution of open and closed channels for $J = 1.25\,k_b T$, $\rho = 0.25$. On the *right* is the histogram of channels with a given number of occupied neighbors, for both open and closed channels (*red*) and only open channels, for different values of τ. On the *left* are shown snapshots of simulations of the system in equilibrium for a range of tensions τ, with closed channels in *black* and open in *grey*. **a** The initial state of the system before the tension is applied, (corresponds to the red distribution in the histogram); **b** System in equilibrium for $\tau = 0.5\,k_b T/\text{nm}^2$ (*grey* distribution in histogram); **c** $\tau = 1.5\,k_b T/\text{nm}^2$ (*cyan* distribution in histogram); **d** $\tau = 2.5\,k_b T/\text{nm}^2$ (*blue* distribution in histogram)

In the case of more compact clusters, obtained for higher values of the interaction parameter ($J = 1.25$), the sites in the "bulk" dominate, and the transition for $k_i = 4$ will be the only significant one (see Fig. 3.20). Nevertheless, we note that for some values of tension, it is still possible to observe coexistence of channels in both conformations.

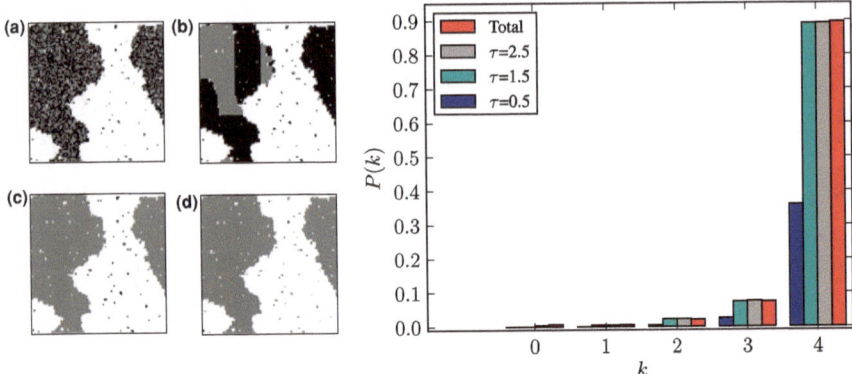

Fig. 3.19 Spatial distribution of open and closed channels for $J = 1.25\,k_bT$, $\rho = 0.5$. On the *right* is the histogram of channels with a given number of occupied neighbors, for both open and closed channels (*red*) and only open channels, for different values of τ. On the *left* are shown snapshots of simulations of the system in equilibrium for a range of tensions τ, with closed channels in *black* and open in *grey*. **a** The initial state of the system before the tension is applied, (corresponds to the *red* distribution in the histogram); **b** System in equilibrium for $\tau = 0.5\,k_bT/nm^2$ (*grey* distribution in histogram); **c** $\tau = 1.5\,k_bT/nm^2$ (*cyan* distribution in histogram); **d** $\tau = 2.5\,k_bT/nm^2$ (*blue* distribution in histogram)

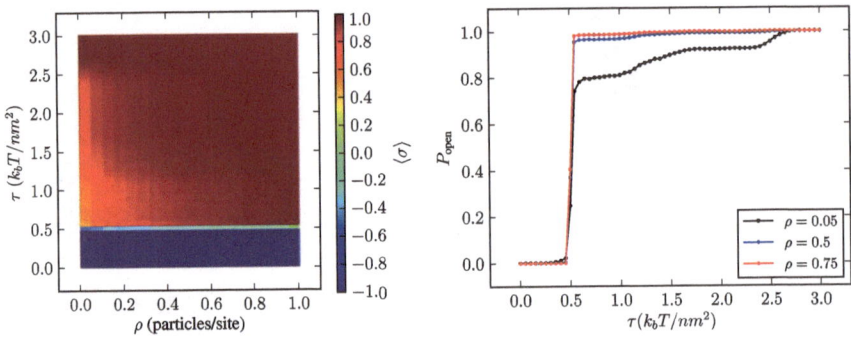

Fig. 3.20 The average channel conformation $\langle\sigma\rangle$ in equilibrium as a function of τ. On the *right* is shown the gating probability P_{open} as a function of the tension τ. The particle spatial distribution is given by the lattice gas with an interaction strength $J = 1.25\,k_bT$

The equilibrium configurations described in this section show how strongly the channel interactions influence the collective gating of mechanosensitive channels. However, we must emphasise that a more biologically realistic situation is not of thermodynamic equilibrium, but instead a configuration shortly after the tension has been abruptly raised, where all channels are initially in a closed state. This situation, as will be explained in the following section, represents a metastable state, which behaves as an equilibrium state form at short time scales, but is significantly different from the true equilibrium states.

3.9.2 Dynamics of Escape From the Metastable State

In physiological conditions the channels diffuse in a closed state on the membrane, when they can agglomerate in clusters. If the tension of the membrane increases due to osmotic shock the channels must first escape the configuration where they are all closed. The transition to a global minimum of the free energy, where a macroscopic fraction of the channels are in the open state, involves the system leaving a local minimum of the free energy, where all channels are closed, and temporarily assuming anti-aligned states with respect to their neighbours. States which correspond to a local minimum of the free energy are denominated metastable, since they retain some of the properties of equilibrium states [31]. The system will eventually escape any metastable state and reach equilibrium, at a global minimum of the free energy. However this can take an arbitrarily long time, and in some cases the escape time can be even comparable to the lifetime of the universe [43]. Therefore metastability is characterised by the existence of significantly distinct time-scales: a short time scale, where the system is in an apparent equilibrium state, and a larger time scale, where the system experiences a transition to equilibrium. For our system we will need to estimate this escape time, in order to determine the response of the bacterium to osmotic shock. This escape time will depend on the interaction strength between the channels and also on thermal fluctuations.

First we start with an extreme case of zero temperature. In this situation the time to leave the metastable state will be infinite, unless the external field is strong enough. We consider for simplicity the case of an entirely occupied lattice ($\rho = 1$). This system is identical to the Ising model, and as expected we observe an hysteresis loop for transitions $-1 \rightarrow +1$ and $1 \rightarrow -1$ (see Fig. 3.21). The presence of additional field P, due to the asymmetries of the energy states, only makes this loop broader, compared to the traditional Ising model. As the temperature increases from 0, this loop narrows, approaching the equilibrium curve (see Fig. 3.21). For the case of lower densities, the system still preserves the hysteresis loop, although it will not be symmetric (see Fig. 3.21). Apart form the isolated channels that gate in the tension $2.5\,k_bT/\text{nm}^2$ we have a major transition to an open conformation at $3\,k_bT/\text{nm}^2$ and to a closed conformation at $1\,k_bT/\text{nm}^2$. The loop is not as broad as in the case of $\rho = 1$, because of the influence of the borders of the clusters.

In the next part we follow with the analysis of the importance of the thermal fluctuations to escape from the metastable state.

3.9.3 The Transition From Closed to Open Conformations

Here we analyse a more realistic situation with the presence of thermal fluctuations. Starting from the metastable state where all channels are closed, we evolve the system for a specific time. First we evolve it for 10^3 Monte-Carlo steps per channel (the right Figs. 3.22 and 3.23). Then we follow with a longer time of 10^5 Monte-Carlo steps

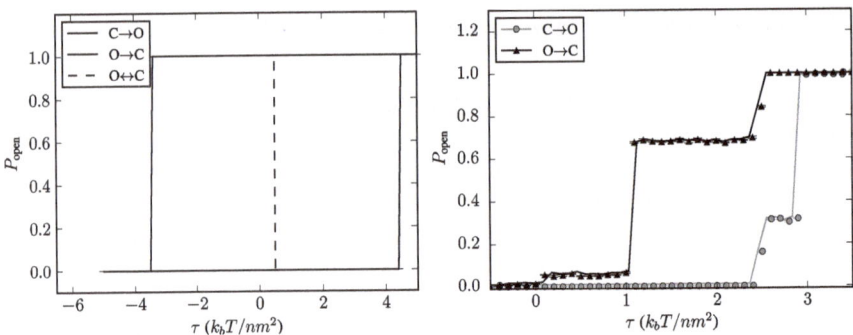

Fig. 3.21 The gating probability P_{open} for $T = 0$, as a function of tension τ, for an initial condition with all channels closed ($C \rightarrow O$), all channels open ($O \rightarrow C$) and channels with random states and $T > 0$ ($O \leftrightarrow C$). On the *left* we consider a full lattice $\rho = 1$, and on the *right* the spatial distribution is given by $\rho = 0.25$ and $J = 1.25$

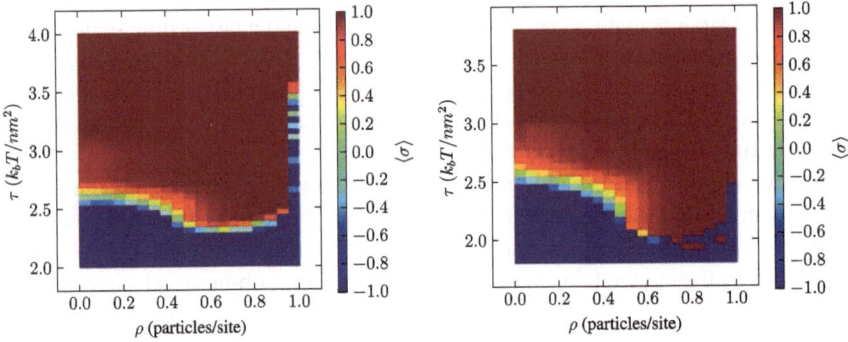

Fig. 3.22 The average channel conformation $\langle \sigma \rangle$ as a function of the tension τ and density ρ, for $J = 0.75\, k_b T$, and an initial metastable configuration where all the channels are closed ($\langle \sigma \rangle = -1$). The figure on the *left* shows the system after 10^3 Monte Carlo steps, and on the *right* after 10^5 Monte Carlo steps

(the left Figs. 3.22 and 3.23). The transitions in both of these cases do not depend as strongly on the form of the clusters (either ramified, with $J = 0.75$ or compact, with $J = 1.25$), compared to the equilibrium case, as Figs. 3.22 and 3.23 show. However it is possible to observe that the system changes its properties for higher densities of $\rho \gtrsim 0.5$ and $\rho \approx 1$, see Figs. 3.22 and 3.23. The value of $\rho = 0.5$ appears to be the turning point for two different types of transitions. Below this point the particles are sparsely distributed and the cluster sizes are not constrained by the finite size of the grid. Above this point the particles are dense enough so they percolate the grid and in these cases the holes (empty lattices) are distributed in finite size aggregates in the space full of particles. Therefore for $\rho \lesssim 0.5$ the global change to equilibrium is induced by the cluster borders (channels with $k_i \in \{1, 2\}$), the transition therefore occurs for the tensions above the gating threshold of isolated channels. However, for $\rho \gtrsim 0.5$ the average cluster size is comparable to the system size, since a value of

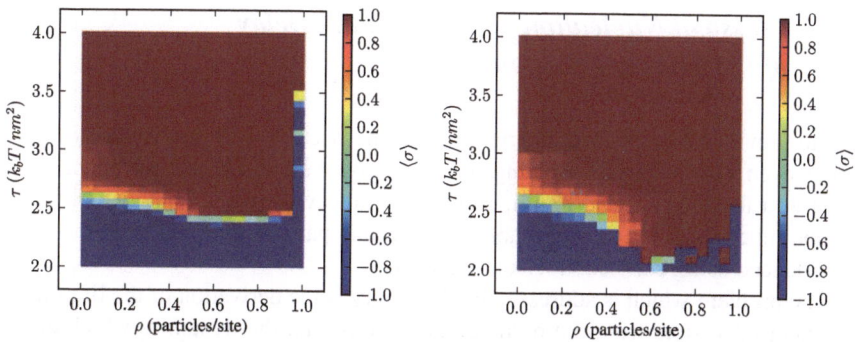

Fig. 3.23 The average channel conformation $\langle \sigma \rangle$ as a function of the tension τ and density ρ, for $J = 1.25 \, k_b T$, and an initial metastable configuration where all the channels are closed ($\langle \sigma \rangle = -1$). The figure on the *left* shows the system after 10^3 Monte Carlo steps, and on the *right* after 10^5 Monte Carlo steps

$\rho = 0.5$ is exactly the percolation threshold where randomly placed channels already lead to clusters spanning the whole system [36]. In this situation the cluster borders are dominated by the bulk, which decrease the gating threshold and the transition is strongly abrupt.

3.9.4 The Transition From Open to Closed Conformations

Here we examine the reverse situation of the previous section: we start with all the channels in the open conformation and analyse the transition to the closed state. This corresponds to the physical situation where the channels must close, after the normalization of the membrane tension. In this case the cluster ramification has a very strong effect, and can be seen in 24, comparing the ramified clusters ($J = 0.75 \, k_b T$) on the left side and the compact clusters ($J = 1.25 \, k_b T$) on the right. For ramified clusters the difference to the individual gating behaviour starts to appear only for high density regions. However for dense clusters the channels stay open after the normal osmolarity is restored.

In this case, similarly to the transition from closed to open state, we can notice a strong difference between the channel's reaction in high and low densities. Although here the turning point is not that obvious as previously, we can speculate that the percolation of the channels on the grid at $\rho = 0.5$ can change the gating dynamics. Therefore the explanation of this behaviour is analogous to the previously described metastable dynamics, where below the percolation point the channel's closure is triggered by the border of the cluster and above it the core of the clusters is large enough to hold the channels in the open state.

3.9.5 Classical Nucleation Theory and the Delay of the Channel Response

In the previous sections we determined that the bacteria will survive the osmotic shock even with its channels acting cooperatively. We observe that the channels do open at a tension of $2.5\,k_bT/\text{nm}$, as desired, however with a delay, which is the escape time of the initial metastable state. In this part we will calculate the delay experienced using classical nucleation theory.

The problem which is addressed by the theory is the derivation of the lifetime of the metastable state from the particle interactions, and the dynamics which drives the escape process. The well established model in the literature to derive the escape time is the droplet model. Central to this model is the concept of a droplet, which is defined as a cluster of particles that have the same state, but are surrounded by others in the opposite state. The droplet can grow or shrink depending on the contribution of the droplet to the free energy. The interior of the droplet contributes to a decrease of the free energy, and the border to an increase. There is a critical droplet size, which represents a transitional barrier, which, when crossed, allows the system to leave the metastable state. The key point of the model is the relation between the lifetime of the metastable state τ and the free energy of the critical droplet F_c, which is given as

$$\langle \tau \rangle \sim e^{F_c/k_bT}. \tag{3.30}$$

The main objective is therefore to obtain this free energy. One approximate but very simple calculation can be done for the Ising model, by assuming that $h < J$, where h is the external field and J the interaction between particles. If we consider the formation of a square droplet of size $l \times l$ inside the metastable state, it has an energy

$$E(l) = 4lJ - hl^2. \tag{3.31}$$

Differentiating the equation we obtain the $E(l)$ has a maximum for

$$l_c = \frac{2J}{h}, \tag{3.32}$$

which is the critical droplet size. Inserting this in Eq. 3.31 we obtain the metastable lifetime as

$$\langle \tau \rangle \sim e^{\beta} \left(\frac{4J'^2}{h'} \right) \tag{3.33}$$

From this result we can estimate the time delay of the gating of mechanosensitive channels for system with $\rho = 1$. Using the approximation just derived we estimate the response time to be

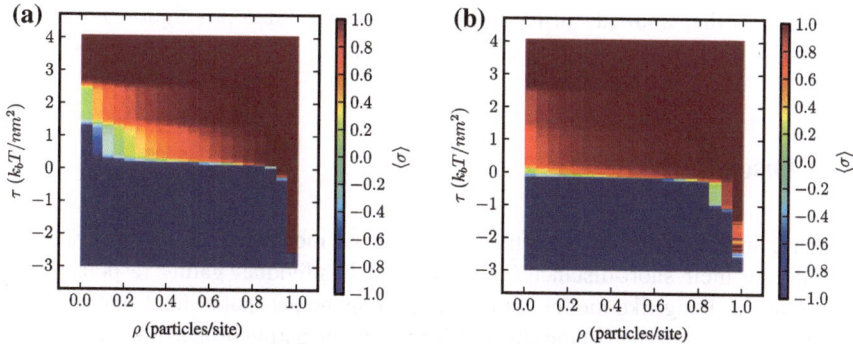

Fig. 3.24 The average channel conformation $\langle \sigma \rangle$ as a function of the tension τ and density ρ, and an initial metastable configuration where all the channels are closed ($\langle \sigma \rangle = +1$), after 10^3 Monte Carlo steps. The figure on the *left* shows the system with $J = 0.75\,k_bT$, and on the *right* with $J = 1.25\,k_bT$

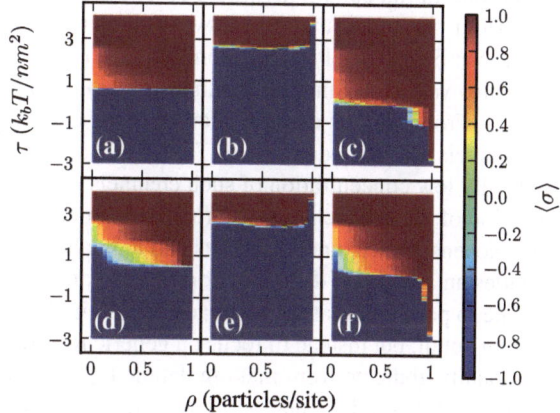

Fig. 3.25 The average channel conformation $\langle \sigma \rangle$ as a function of the tension τ and density ρ, for both the equilibrium states and the escape from the metastable states. The *top* row shows the properties of a spatial configuration where dense clusters are formed ($J = 1.25\,k_bT$). On the *bottom* row are represented the properties of the system characterised by the formation of ramified clusters ($J = 0.75k_bT$). **a, d** System in equilibrium; **b, e** Escape form the initial metastable state corresponding to all channels closed; **c, f** Escape from the metastable state corresponding to all channels open

$$\langle \tau \rangle \sim e^{\frac{2E^2}{k_bT\,(2P-h)}} \tag{3.34}$$

Monte-Carlo steps per channel (mcs/c). Assuming that each mcs/c corresponds to the characteristic reaction time of a single channel, i.e., $\sim 3\,\mu s$, this gives us a gating response of ~ 20 ms (Fig. 3.24).

We summarise the results obtained with our approach in Fig. 3.25, where we can see how the gating process changes for the case of the dense compact clusters compared to the ramified cluster spatial distribution. Also this figure shows how different is the metastable gating dynamics to the gating of the equilibrium system.

3.10 Discussion

We have investigated the collective behaviour of mechanosensitive channels, by considering their short-distance interactions and individual gating responses. We elaborated an Ising-like model with a non-homogeneous spatial field. We analysed with this model the effect of the channel density on the gating properties. We obtained that properties of the system change strongly for different spatial distributions of the channels. In particular, we observed that for equilibrium the channel clustering would cause leaking in physiological tensions. Some of the channels in the core of the clusters have a tendency to open for tensions as low as $0.5\,k_bT/\text{nm}^2$.

In biologically relevant initial conditions where all channels are closed, the transition to the equilibrium state is extremely slow, leading to a response time to osmotic shock which can be extremely long. Moreover, the reverse transition back to the closed state is also very slow, and the channels remain open even for already normalized conditions. From this we can speculate that the collective response of mechanosensitive channels is largely detrimental for the cell survival, and it may be a reason for the typically low concentration of such channels on the cell membrane. However, if the function of the channels is not to regulate osmotic stress, such as the other types of mechanosensitive channels in plants and animals, it could be the case that clustering provides an additional degree of control to channel gating.

Here we would like to point out that the approach chosen in this chapter is not to model the system in all detail, but instead to focus on generic features of this system. Therefore we intentionally abdicate from making detailed quantitative predictions, in favour of isolating the most fundamental properties which are responsible for the most relevant aspects of the system. This allowed us to capture fundamental emergent properties of the system, which can in principle be easily verified with the appropriate experimental methods. We described some of necessary methods in Sect. 3.3 of this chapter, page 36. The verification would involve methods for spatial localisation together with a patch-clamp to measure the gating. For even more realistic results, perhaps it would be more appropriate to extend the analysis to lattices with different shapes. Such as for example an hexagonal lattice. Since the gating is influenced by the spatial organization, it would be interesting to investigate the effects of more detailed space geometries. Additionally, in our approach it was very important to consider the gating process in a different time scale of the diffusion rate. However it is possible to combine the two time scales, similar to what was done in the Blume–Emery–Griffiths model [44], and consider both the diffusion and the conformation change simultaneously.

Furthermore, although we focus on a specific bacterial channel, our approach can be extrapolated to other channels that are affected by physical properties of the membrane in a similar way. The observed properties are the direct result of the nature of the interacting forces, and should not be strongly affected by any neglected secondary detail. For instance, for the mechanosensitive channels of small conductance of *E. coli* our model could be used with only minor modifications. As we described in Sect. 3.6.2.3, page 65, these channels not only deform the thickness of the membrane, which results in a short-range attraction or repulsion, but also exert a midplane deformation, generating a repulsive barrier for larger distances. Also, it would be interesting to investigate the interaction between mechanosensitive channels of large and small conductance together. These investigations would give a more complete understanding of the osmotic response in bacteria.

It is also possible to extend this approach even further to proteins two (or more) different active states. In the review [45] the authors list several membrane proteins with available structural information, which have more then one conformation. Examples are proteins such as KcsA (1BL8, 1JQ2), NaK channel (2AHY, 3E86), bacteriorhodopsin (1C8R, 1C8S) [45]. Additionally, the approach could in principle be extended to other types of mechanosensitive channels such as the channels of Eukaryotes, described in Sect. 3.2.2, page 35.

Finally, the fact that this system can be mapped elegantly into an new class of spin model—composed of an Ising model with a spatially-correlated local field, which is a result of a lattice gas equilibrium configuration—serves as motivation to the detailed study of this new class of model, and thus is also an important contribution to the physics of biologically inspired systems.

References

1. Destainville, N.: Cluster phases of membrane proteins. Phys. Rev. E **77**(1), 011905 (2008)
2. Booth, I.R., Edwards, M.D., Black, S., Schumann, U., Miller, S.: Mechanosensitive channels in bacteria: signs of closure? Nat. Rev. Microbiol. **5**(6), 431–440 (2007)
3. Guseva, K., Thiel, M., Booth, I., Miller, S., Grebogi, C., de Moura, A.: Collective response of self-organised clusters of mechanosensitive channels. Phys. Rev. E Stat. Nonlin. Soft. Matter. Phys. **18**, 020901 (2011)
4. Kung, C.: A possible unifying principle for mechanosensation. Nature **436**(7051), 647–654 (2005)
5. Hamill, O.P.: Mechanosensitive Ion Channels. Academic Press, New York (2007)
6. Kamkin, A.: Mechanosensitive Ion Channels, 1st edn. Springer, Heidelberg (2007)
7. Arnadóttir, J., Chalfie, M.: Eukaryotic mechanosensitive channels. Annu. Rev. Biophys. **39**, 111–137 (2010)
8. Hamill, O.P., Martinac, B.: Molecular basis of mechanotransduction in living cells. Physiol. Rev. **81**(2), 685–740 (2001)
9. Phillips, R., Ursell, T., Wiggins, P., Sens, P.: Emerging roles for lipids in shaping membrane-protein function. Nature **459**(7245) 379–385 (2009)
10. Chang, G., Spencer, R.H., Lee, A.T., Barclay, M.T., Rees, D.C.: Structure of the MscL homolog from *Mycobacterium tuberculosis*: a gated mechanosensitive ion channel. Science **282**(5397), 2220–2226 (1998)
11. Sukharev, S., Betanzos, M., Chiang, C.-S., Robert Guy, H.: The gating mechanism of the large mechanosensitive channel MscL. Nature **409**(6821), 720–724 (2001)

12. Sukharev, S., Durell, S.R., Guy, H.R.: Structural models of the MscL gating mechanism. Biophys. J. **81**(2), 917–936 (2001)
13. Cruickshank, C.: Estimation of the pore size of the large-conductance mechanosensitive ion channel of *Escherichia coli*. Biophys. J. **73**(4), 1925–1931 (1997)
14. Sukharev, S.I., Martinac, B., Arshavsky, V.Y., Kung, C.: Two types of mechanosensitive channels in the *Escherichia coli* cell envelope: solubilization and functional reconstitution. Biophys. J. **65**(1), 177–183 (1993)
15. Perozo, E.: Gating prokaryotic mechanosensitive channels. Nat. Rev. Mol. Cell. Biol. **7**(2), 109–119 (2006)
16. Bass, R.B., Strop, P., Barclay, M., Rees, D.C.: Crystal structure of *Escherichia coli* MscS a voltage-modulated and mechanosensitive channel. Science **298**(5598), 1582–1587 (2002)
17. Wang, W., Black, S.S., Edwards, M.D., Miller, S., Morrison, E.L., Bartlett, W., Dong, C., Naismith, J.H., Booth, I.R.: The structure of an open form of an *E. coli* mechanosensitive channel at 3.45 Å resolution. Science **321**(5893), 1179–1183 (2008)
18. Powl, A.M., East, J.M., Lee, A.G.: Lipid–protein interactions studied by introduction of a tryptophan residue: the mechanosensitive channel MscL. Biochem. **42**(48), 14306–14317 (2003)
19. Perozo, E., Kloda, A., Marien Cortes, D., Martinac, B.: Physical principles underlying the transduction of bilayer deformation forces during mechanosensitive channel gating. Nat. Struct. Mol. Biol. **9**(9), 696–703 (2002)
20. Martinac, B., Buechner, M., Delcour, A.H., Adler, J., Kung, C.: Pressure-sensitive ion channel in *Escherichia coli*. Proc. Nat. Acad. Sci. U. S. A. **84**(8), 2297–2301 (1987)
21. Norman, C., Liu, Z.-.W., Rigby, P., Raso, A., Petrov, Y., Martinac, B.: Visualisation of the mechanosensitive channel of large conductance in bacteria using confocal microscopy. Eur. Biophys. J. **34**(5), 396–402 (2005)
22. Haswell, E.S., Meyerowitz, E.M.: MscS-like proteins control plastid size and shape in *Arabidopsis thaliana*. Curr. Biol. **16**(1), 1–11 (2006)
23. Zhang, S., Arnadottir, J., Keller, C., Caldwell, G.A., Yao, C. A., Chalfie, M.: MEC-2 is recruited to the putative mechanosensory complex in *C. elegans* touch receptor neurons through its stomatin-like domain. Curr. Biol. **14**(21), 1888–1896 (2004)
24. Ursell, T., Huang, K.C., Peterson, E. Phillips, R.: Cooperative gating and spatial organization of membrane proteins through elastic interactions. PLoS Comput. Biol. **3**(5), e81 (2007)
25. Jackson, J.D.: Classical Electrodynamics, 3rd edn. Wiley, New York (1998)
26. Weiss, T.F.: Cellular Biophysics, vol. 1, Transport. The MIT Press, Cambridge (1996)
27. Casimir, H.B.G., Polder, D.: The influence of retardation on the London-van der waals forces. Phys. Rev. **73**(4), 360 (1948)
28. Goulian, M., Bruinsma, R., Pincus, P.: Long-range forces in heterogeneous fluid membranes. Europhys. Lett. (EPL) **22**(2), 145–150 (1993)
29. Bruinsma, R., Pincus, P.: Protein aggregation in membranes. Curr. Opin. Solid State Mater. Sci. **1**(3), 401–406 (1996)
30. Lipowsky, R., Sackmann, E.: Structure and Dynamics of Membranes. Elsevier, Amsterdam (1995)
31. Landau L.D., Lifshitz E.M.: Statistical Physics, part 1, vol. 5, 3rd edn. Butterworth-Heinemann, Oxford (1980)
32. Marrink, S.J., de Vries, A.H., Mark, A.E.: Coarse grained model for semiquantitative lipid simulations. J. Phys. Chem. B **108**(2), 750–760 (2004)
33. West, B., Brown, F., Schmid, F.: Membrane-protein interactions in a generic coarse-grained model for lipid bilayers. Biophys. J. **96**(1), 101–115 (2009)
34. de Meyer, F. J.-M, Venturoli, M., Smit, B.: Molecular simulations of lipid-mediated protein–protein interactions. Biophys. J. **95**(4), 1851–1865 (2008)
35. Landau L.D., Lifshitz E.M.: Mechanics, vol. 1, 3rd edn. Butterworth-Heinemann, Oxford (1976)
36. Newman, M.E.J., Barkema G.T.: Monte Carlo Methods in Statistical Physics. Oxford University Press, Oxford (1999)

37. Binney, J.J., Dowrick, N.J., Fisher, A.J., Newman, M.E.J.: The Theory of Critical Phenomena: An Introduction to the Renormalization Group. Oxford University Press, USA (1992)
38. Onsager, L.: Crystal statistics. I. A two-dimensional model with an order-disorder transition. Phys. Rev. **65**(3–4), 117 (1944)
39. Phillips, R., Kondev, J., Theriot, J.: Physical Biology of the Cell. Garland Science, New York (2008)
40. Shapovalov, G., Lester, H.A.: Gating transitions in bacterial ion channels measured at 3 microns resolution. J. Gen. Physiol **124**(2), 151–161 (2004)
41. Ramadurai, S., Holt, V.K.A., van den Bogaart, G., Killian, J.A., Poolman, B.: Lateral diffusion of membrane proteins. J. Am. Chem. Soc. **131**(35), 12650–12656 (2009)
42. Metropolis, N., Rosenbluth, A.W., Rosenbluth, M.N., Teller, A.H., Teller, E.: Equation of state calculations by fast computing machines. J. Chem. Phys. **21**(6), 1087–1092 (1953)
43. Rikvold, P.A., Tomita, H., Miyashita, S., Sides, S.W.: Metastable lifetimes in a kinetic Ising model: dependence on field and system size. Phys. Rev. E **49**(6):5080 (1994)
44. Blume, M., Emery, V.J., Griffiths, R.B.: Ising model for the lambda transition and phase separation in He^3–He^4 mixtures. Phys. Rev. A **4**(3):1071 (1971)
45. Lundbæk, J.A., Collingwood, S.A., Ingólfsson, H.I., Kapoor, R., Andersen, O.S.: Lipid bilayer regulation of membrane protein function: gramicidin channels as molecular force probes. J. R. Soc. Interface **7**(44), 373–395 (2010)

Chapter 4
Assembly and Fragmentation of Tat Pores

4.1 Introduction

In this chapter we will analyse how the cell deliberately uses aggregation to construct temporary pores on the membrane. The transient character of these pores characterizes the main feature in the assembly dynamics. We already analysed one form of assembly process and the importance of fragmentation in the Chap. 1. However that process was characterised by a fragmentation rate which is much slower then the rate of assembly. In this chapter, we will analyse a process where the fragmentation is very fast and only proceeds when the assembly achieves a certain limit. This type of dynamics is the main property of the system responsible for a controlled export of folded proteins out of the cell without damaging the cell membrane and avoiding unnecessary loss of intracellular components: The transient pores are only present for enough time to allow the transport of protein, and after the transport is completed the pore is sealed.

We start with a short biological introduction describing the main characteristics of the system in the Sect. 4.2. We follow with a simple theory of single ring formation on the membrane based on the first-passage time on a sphere, described in Chap. 2. Section 4.4.2 brings more complex aspects of the dynamics, where we construct a theory of assembly and fragmentation of the rings of diverse sizes on the membrane. The objective of the last part is, based on a theory for the assembly processes, to establish how this dynamics influences the protein out-flux. Our main result is a quadratic relation between protein size and the maximum rate of efflux.

4.2 Tat Protein Transport System

The protein transport across the lipid bilayer is vital for many organisms. In bacteria and chloroplasts it has a very similar form and is performed by two specially adapted translocases. The first one transports the proteins when they are still unfolded, and is

K. Guseva, *Formation and Cooperative Behaviour of Protein Complexes on the Cell Membrane*, Springer Theses, DOI: 10.1007/978-3-642-23988-5_4, © Springer-Verlag Berlin Heidelberg 2012

called "Sec translocase". The second one takes place when the proteins are already folded, and it is called "twin arginine translocation system (Tat)". Although the Tat system is responsible for the transport of a smaller number of proteins, it is vital for bacteria. It is involved in the transport of proteins responsible for energy metabolism, formation of the cell envelope, biofilm formation, heavy metal resistance, nitrogen-fixing symbiosis, and also bacterial pathogens [1]. The mechanisms in the core of this process is also more complex since it has to allow passage of very big proteins of various sizes in a very selective way without leaking other vital cell constituents. This is done by the construction of transient pores, which is triggered by a signal given by the system's substrate proteins and it varies in size, according to the size of the protein to be transported [1]. After the protein is transported the pore is sealed.

The Tat translocation system in *E. coli* is composed by three proteins: TatA, TatB and TatC. All of them are membrane proteins and form dynamic complexes. The TatB and TatC form the TatBC receptor complex that, after binding to the substrate protein, engages the transport process, which involves the polymerization of TatA subunits into pores for the protein transport. The size of the assembled pores are heterogeneous and correspond to the diameter of the transported protein. The mechanism of the protein translocation trough the pore is not well understood. However, the current models consider the protomotive force being the main component that drives the process [2].

Although it is understood that the TatA oligomerisation is the core of the transport dynamics, there are still not many detailed studies of this process. However it is generally accepted that it takes place according to the following steps:

1. *Production of protein inside the cell.* There are several proteins produced inside the cell which are destined to be exported. For as long as they are produced they diffuse and reach the membrane from the cytosol. They have an N-terminal domain that signals a request for translocation.
2. *Pore formation.* The TatA subunits assemble forming a ring of the exact size necessary for the transport of the protein (see Fig. 4.1).
3. *Protein translocation.* After the pore ring is assembled, the protein is forced to leave the cell.
4. *Pore fragmentation.* After the protein is transported trough the pore, the ring subunits disassemble back to their free form.

Studies on the stochiometry of the Tat complexes show that they are composed on average of 25 subunits [2]. Also in the same study there are estimations of a number of ∼100 free TatA molecules dispersed on the membrane, and on average 15 rings in the process of assembly. In the next section we construct a theoretical model based on the information described in this section. We start relating the assembly process with the time of transport of a single protein. Then we follow with a theory that relates the protein outflux and the dynamics of subunit assembly.

Fig. 4.1 The oligomerisation process of TatA rings

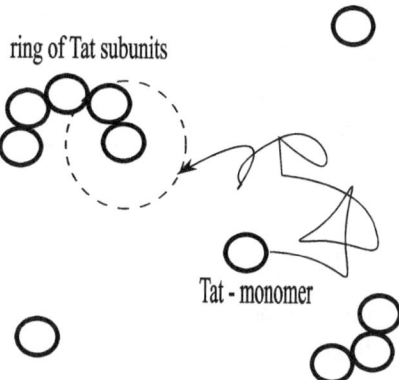

ring of Tat subunits

Tat - monomer

4.3 The Theory of TatA Assembly Process

In this part we elaborate a quantitative model for the process of pore assembly. We assume the following conditions:

1. *We describe only the TatA oligomerisation*. As we described previously, the Tat system is composed by three proteins TatA, TatB and TatC. However, both TatB and TatC are only important in the initiation step of the process, and the pore assembly steps are only dependent on TatA. It has been noted that "the oligomeric state of the TatA is the key to understand the translocation" [2]. Since we are interested in the influence of the dynamics of aggregation on the transport process our theoretical approach takes in to account only the TatA subunits.

2. *TatA as a monomer*. What we call TatA monomers are actually constituted by four subunits aggregated into a small complex. However, since this is the way these proteins are usually found on the membrane we, use this small complex as a unit of mass.

3. *Free diffusion*. We consider that the TatA monomer diffuse freely on the membrane. This is a type of mean-field approximation, where spatial correlations are neglected.

4. *TatA form rings*. The TatA monomers diffuses until it finds the end of an existing TatA ring, or a signal to initiate ring formation. Assuming the TatA's have to attach to the end of an existing ring, a TatA monomer must come within a certain distance l of the TatA at the end of the ring to add to it, where l is of the order of the diameter of a TatA monomer. We assume that this process works as a perfect trap for TatA, meaning that the monomers are always incorporated in to a ring, if within the distance l around it. The rings grow only by single monomer assembly and can not fuse with each other as in the process described in Chap. 2.

5. *Fragmentation*. The formed aggregates can fragment back into monomers. However the fragmentation only occurs for complexes after the translocation of the protein is completed. Therefore, the monomers disassociate after the complex reaches predefined sizes.

Table 4.1 Parameters for TatA and the bacteria cell used in the estimation of the rate of attachment of new monomers to an initiated ring, according to Ref. [2]

Cell radius	R	$\sim 10^{-6}\,\text{m}$
TatA subunit radius	r	$\sim 10^{-9}\,\text{m}$
TatA diffusion coefficient	D	$\sim 0.13 \times 10^{-12}\,\text{m}^2/s$

6. *Time scale.* We consider that the assembly of the subunits is the main process defining the time scale of protein transport. In other words, we consider that the time of protein transduction trough a formed pore, and the following time of pore fragmentation are irrelevant when compared with the time of subunit assembly.

To calculate the average time for a single TatA monomer to be incorporated into an initiated ring we use the first passage time approach explained previously in Sect. 2.3, Chap. 2, page 11. Assuming this process works as a perfect trap for TatA's, the average time $\langle t \rangle$ it takes for a monomer starting from a random position on a sphere of radius R to reach a ring is

$$\langle t \rangle = \frac{2R^2}{D} \ln\left(\frac{R}{r}\right). \tag{4.1}$$

where D is the diffusion coefficient of a TatA monomer. The estimation of for the diffusion of a TatA monomer is $D \sim 0.13\ 10^{-12}\,\text{m}^2/s$ [2, 3], see Table 4.1. Using the parameters from Table 4.1 and Eq. 4.1 we estimate $\langle t \rangle \sim 100$ s.

4.3.1 The Assembly of a Single Ring

In this part we shortly describe the formation of a single ring of TatA subunits in a pool of TatA monomers. We start with a system with a sea of free TatA subunits diffusing freely, and a single ring initiating point. The number of free TatA subunits n_1 as a function of time can be described by

$$\frac{dn_1(t)}{dt} = -\frac{n_1}{\langle t \rangle}, \tag{4.2}$$

Therefore the formation of a single channel of n subunits will be described by

$$n(t) = N_{\text{free}}(1 - e^{-t/\langle t \rangle}), \tag{4.3}$$

where n is the number of particles assembled into a channel, and $N_{\text{free}} = n_1(t = 0)$ is the initial number of free monomers on the membrane. Therefore we can estimate the time t_n of formation of a ring of size n as

$$t_n = \langle t \rangle \ln\left(\frac{N_{\text{free}}}{N_{\text{free}} - n}\right) \tag{4.4}$$

Fig. 4.2 Time of formation
t_n of a ring n monomers as a
function of n, for
monodisperse initial
conditions with N_{free}
monomers

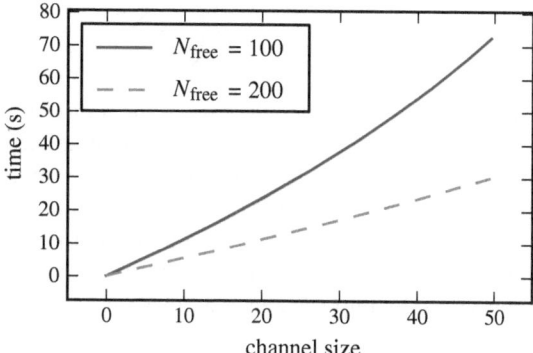

From this calculation we obtained that the time to form an average channel size of 25 monomers is around 30 s for an initial condition of $N_{\text{free}} = 100$. An initial condition with larger number of monomers decreases this time, see Fig. 4.2.

In the real situation, however, there are many rings being formed and decomposed on the membrane as long as the protein flux to the periplasm is in progress. Therefore the next sections are dedicated to this dynamics.

4.4 Assembly of Multiple Rings

In this part we explain the complex dynamics of formation and sealing of the pores on the membrane surface required for protein transport. We describe the system in terms of two main parameters: the number of initiated but incomplete rings of final size i on the membrane, N_i, and the number of free monomers on the membrane, n.

We can consider N_i to be the number of proteins which require a ring of i TatA monomers around them to be exported, and which have not been exported yet. They all have incomplete TatA rings of various lengths around them. When a ring is completed, one of the N_i proteins disappears from the membrane and is exported into the periplasm; this causes N_i to decrease by 1. There is also an influx of ϕ_i size-i proteins into the membrane coming from the cytoplasm. The dynamics of N_i can be described by

$$\frac{dN_i}{dt} = -k_i(n, \{N_j\})N_i + \phi_i. \tag{4.5}$$

Here k_i measures the fragmentation of i-sized proteins caused by ring completion. Therefore, k_i is not a constant, and we must determine it eventually in order to use Eq. 4.5.

The differential equations satisfied by the number of free monomers n is

$$\frac{dn}{dt} = -\left(\frac{1}{\langle t \rangle} \sum_i N_i\right) n + \sum_i i k_i N_i. \qquad (4.6)$$

The first term represents the loss of monomers to growing rings. The second term comes from the fact that when a ring is completed, the protein is exported and the monomers are released back to the pool of free monomers. A ring of size i fragments into i monomers, hence the factor i in the sum.

Another important feature of this system is that the total number of monomers must be conserved, since there is no additional production of TatA. This can be represented by imposing an additional constraint

$$M = n(t) + \kappa \sum_i i N_i(t), \qquad (4.7)$$

where κ represents the average number of monomers in a ring. We follow by the analysis of the steady state properties of the system just described.

4.4.1 Translocation of Only One Type of Protein

We start with a slight simplification of the system which consists in the analysis of the translocation of only one type of protein of size m. For this case, the system of Eqs. 4.5, 4.6 and 4.7 can be written as a single differential equation for the number of free monomers

$$\frac{dn}{dt} = \frac{n^2}{\langle t \rangle \kappa m} - \frac{Mn}{\langle t \rangle \kappa m} + m\phi_m. \qquad (4.8)$$

This differential equation has two fixed points for

$$n = \frac{M \pm \sqrt{M^2 - 4\langle t \rangle \kappa m^2 \phi_m}}{2} \quad \Rightarrow \quad N_m = \frac{M \mp \sqrt{M^2 - 4\langle t \rangle \kappa m^2 \phi_m}}{m} \qquad (4.9)$$

where both solutions are positive and satisfy $n < M$. The fixed point with the larger number of monomers is the stable fixed point the other one is unstable (see Fig. 4.3). Thus, in a real system we would expect to find the situation with more free monomers and less assembled complexes.

4.4.2 Translocation of Protein of Distinct Sizes

Assuming the system reaches a steady state, the total number of free TatA monomers n and the number of initiated rings N_i on the membrane are constant. This implies that there should be a constant release of monomers into the pool as well as a constant

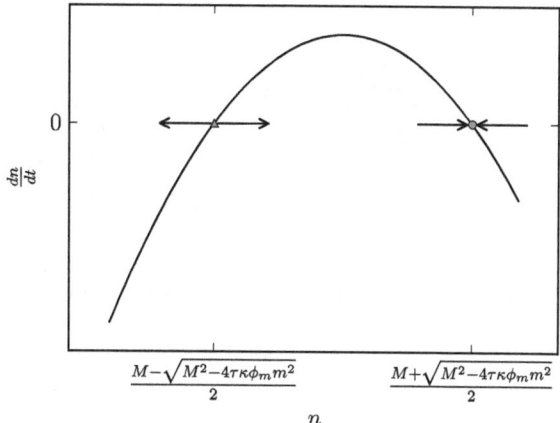

Fig. 4.3 The phase space diagram of Eq. 4.9, showing the stable (*circle*) and unstable (*triangle*) fixed points of the system

monomer uptake to form the rings. We denominate the flux of monomers released by Φ which is equivalent to

$$\Phi = \sum_i i\phi_i = \sum_i ik_i N_i. \tag{4.10}$$

In this model of TatA aggregation, each monomer can bind to any of the existing rings around the proteins with equal probability. In the steady state, the number of monomers released in 1 s is equal to the number of monomers attaching to rings during that time. So on average each protein is receiving α monomers per second in the steady state, with

$$\alpha = \frac{\Phi}{N}, \tag{4.11}$$

where $N = \sum_i N_i$ is the total number of proteins on the membrane. One monomer represents $1/i$ of a ring of size i, so the rate of completion of a ring around a protein of size i in steady state is

$$k_i = \frac{\alpha}{i} = \frac{\Phi}{iN}. \tag{4.12}$$

Making $\frac{dN_i}{dt} = 0$ in Eq. 4.5, we get $k_i N_i = \phi_i$. If we replace this expression into Eq. 4.12, we find

$$\frac{N}{\Phi} = \frac{N_i}{i\phi_i} = \frac{N_j}{j\phi_j}, \tag{4.13}$$

for any i, j.

Setting $\frac{dn}{dt} = 0$ in Eq. 4.6 and using Eq. 4.12, we find

$$Nn = \Phi \langle t \rangle, \tag{4.14}$$

where we used $\sum_i i k_i N_i = \sum_i \phi_i = \Phi$.

Using Eq. 4.13, the term $\sum_i i N_i$ in Eq. 4.7 can be written as $(N/\Phi) \sum_i i^2 \phi_i = (N/\Phi)\Psi$, where we define

$$\Psi = \sum_i i^2 \phi_i. \tag{4.15}$$

Equation 4.7 can then be rearranged as a relation between N and n:

$$N = (M - n)\frac{\Phi}{\Psi \kappa}. \tag{4.16}$$

Using Eqs. 4.14 and 4.16, we can get a closed equation for n:

$$N = \frac{\Phi \langle t \rangle}{n} = (M - n)\frac{\Phi}{\Psi \kappa} \quad \Rightarrow \quad n^2 - Mn + \kappa \langle t \rangle \Psi = 0. \tag{4.17}$$

This is a quadratic equation for n, with two valid solutions:

$$n = \frac{M \pm \sqrt{M^2 - 4\kappa \langle t \rangle \Psi}}{2} \quad \Rightarrow \quad N_m = \frac{M \mp \sqrt{M^2 - 4 \langle t \rangle \kappa \Psi}}{m} \tag{4.18}$$

Both solutions are positive, and both satisfy $n < M$. The steady state of only on type of particle is in accordance with this solution (see Eq. 4.9).

4.4.3 Condition for the Existence of a Steady State

From Eq. 4.18, the steady state is only possible if

$$M^2 \geq 4\kappa \langle t \rangle \Psi. \tag{4.19}$$

If this condition is violated, proteins are not exported as fast as they are produced, and they will accumulate. So this relation establishes the maximum possible "export rate" which the system can support. The saturation of the Tat translocase has been observed in experiments. These experimental studies also show that it can be relieved only by the overexpression of Tat complex components [4].

The maximum export rate is determined by Ψ, which has interesting consequences. For example, consider the simplest case of having just one protein size m. Then the maximum export rate ϕ_m is, from Eq. 4.19,

$$\phi_m = \frac{1}{4\kappa \langle t \rangle} \left(\frac{M}{m}\right)^2. \tag{4.20}$$

For 100 monomers and a ring of size 20, the maximum transport rate ϕ_m is $25/200$, or a little more that 1 transported molecule every 10 s.

One important prediction is that the maximum rate of transport is inversely proportional to the square of the ring size m. So the maximum rate of transport of a protein which requires a ring of size 40 is four times smaller than that of a protein with a ring of size 20. This can be explained intuitively by the fact that bigger rings take longer because they simply require to be found by more monomers, *and* because while they are being assembled, they sequester more TatA molecules from the free pool, making them unavailable to help assembly of other proteins. These two combined effects are responsible for the rate of transport decreasing faster than linearly with protein size. Conversely, the maximum transport rate also increases with the square of the number of monomers.

4.5 Discussion

We have analysed, in this chapter, how the dynamics of assembly and fragmentation processes on the membrane influences the export of proteins through the membrane by the Tat system. We proposed a description of the system by a set of differential equations which describe the number of free subunits on the membrane and also the number of assembled complexes. From this approach we derived the properties of the system in the steady state. As expected, we observed that the translocation system can be saturated by an excess of demand for transport, as found in experiments [4]. Furthermore we found that the relation between the protein flux and it size is quadratic. In other words, an increase in the size of protein makes its transport disproportionally more difficult.

This is still a work in progress and there are more complex properties to be studied about the system. The work should proceed with the detailed analysis of the influence of stochasticity in this system. For this we should proceed with more extensive numerical simulations. Also we propose the analysis of the simultaneous transport of proteins of distinct sizes and how this influences the dynamics. Furthermore the theory should be extended to contain the whole Tat components, including the receptor TatBC. The receptor has a role in the control of the initiated polimerization, and its limited number on the membrane surface can prevent the saturation of the process. Additionally, the results should be tested with possible experiments. The bacteria *E. coli* doubles the number of proteins in the periplasm using this translocation system, during its life cycle of 20 min. We also propose a quantitative study of the transported proteins and the order in which they are exported to the periplasm. We suggest that this order cannot be arbitrary, and should be controlled for an optimum export rate.

References

1. Gohlke, U., Pullan, L., McDevitt, C.A., Porcelli, I., de Leeuw, E., Palmer, T., Saibil, H.R., Berks, B.C.: The TatA component of the twin-arginine protein transport system forms channel complexes of variable diameter. Proc. Nat. Acad. Sci. U. S. A. **102**(30), 10482–10486 (2005)
2. Leake, M.C., Greene, N.P., Godun, R.M., Granjon, T., Buchanan, G., Chen, S., Berry, R.M., Palmer, T., Berks B.C.: Variable stoichiometry of the TatA component of the twin-arginine protein transport system observed by in vivo single-molecule imaging. Proc. Nat. Acad. Sci. U. S. A. **105**(40), 15376 –15381 (2008)
3. Mullineaux, C.W., Nenninger, A., Ray, N., Robinson C.: Diffusion of green fluorescent protein in three cell environments in *Escherichia coli*. J. Bacteriol. **188**(10), 3442–3448 (2006)
4. Yahr, T.L., Wickner, W.T.: Functional reconstitution of bacterial Tat translocation in vitro. EMBO J. **20**(10), 2472–2479 (2001)

Chapter 5
Conclusion

In this work we studied the dynamics and spatial organization of proteins on prokaryotic membranes. Using theoretical modeling, we show the importance of protein interactions in processes ranging from the formation of protein complexes to transport and homeostasis. We focus on the roles of mutual interactions in the assembly of protein complexes, as well as in the spatial organization and the collective function of assembled complexes on the membrane.

The crystal structure of membrane proteins is particularly difficult to be determined due to the experimental difficulties of obtaining good crystals. However, here we have shown that this information can be very important to understand the organisation of the membrane. Using this information as the core of our theoretical approach, we were able to describe the consequences of protein interactions for the formation of complexes as well as for the collective behaviour of complexes on the membrane. Therefore, the crystal structure can reveal not only the specific function of an isolated protein, but it also provides the necessary information to understand large scale, emerging properties of proteins acting on the membrane.

In the first part of this work we analysed how the interactions among subunits can affect the assembly of protein complexes on the membrane surface. We started from the premise that organisms would try to avoid wastage, and analysed the effect of the interaction strength in the efficiency of the assembly process. The mechanism by which the interaction strength influences the efficiency is by resulting in a fragmentation rate, which reverts the assembly of subunits. We found that lower fragmentation leads to higher efficiency, however for a zero rate of fragmentation, the process becomes again very inefficient. It is important to emphasize that the efficiency achieved through less fragmentation comes at the cost of a slower assembly process. For larger complexes, the optimization of the efficiency can easily lead to assembly time scales which exceed the life cycle of the organism. Therefore, we speculate that this imposes an upper bound on the maximum size of the complexes which can be viably constructed. Also this time restriction results in an optimum regime of fragmentation rate. This suggests that the fragmentation rate should be subjected to evolutionary pressure to be tuned to maximize the final production.

K. Guseva, *Formation and Cooperative Behaviour of Protein Complexes on the Cell Membrane*, Springer Theses, DOI: 10.1007/978-3-642-23988-5_5, © Springer-Verlag Berlin Heidelberg 2012

In the second part of this work, we analysed how membrane-mediated interactions can affect the collective behaviour of assembled complexes on the membrane. As a case study, we considered a special type of channel which opens according to the membrane stretch, called a mechanosensitive channel. These channels are very sensitive to membrane deformations in its surroundings, and can change their conformational state accordingly. This results in a mutual interaction among adjacent channels, which depends on their conformations. By considering the nature and the strength of these interactions, we constructed a coarse-grained model which captures the essential properties of many mechanosensitive channels on the membrane. By analysing the equilibrium and dynamical properties of this system, we observed the formation of clusters of aggregated channels, which behave radically different than their isolated counterparts. These clustered channels exhibit a delayed activation time, as well as a differentiated activation pattern, where channels in the interior of the clusters activate at tensions which are lower than necessary for channels which are at the boundaries. From this results, it was possible to explain why the concentration of such channels is very low in bacteria, since their cooperative behaviour—in particular the resulting delay in activation—would be detrimental for cell survival. However, for other types of mechanosensitive channels, such as those in eukaryotes, this type of cooperative behaviour can be viewed as an additional form of control, and thus be an important part of their function.

In the third, and last part of the work we analysed the mechanism by which cells export proteins form their interior, through the formation of temporary pores on the membrane that allow protein translocation. These pores are formed for a very short period by independent subunit association and always have the exact size of the transported protein. The transported proteins, on the other hand, can have different sizes and need larger or smaller pores. Using differential equation approach, we analytically obtained the properties of the steady state and the maximum export rate as function of protein size and protein flux.

The systems analysed in this work serve as good examples of how the description of the essential properties of protein interactions can provide a fundamental understanding of the emerging properties of collective protein behaviour. As more information on the structure of other membrane proteins are revealed, we propose the appropriate extension of our theoretical approaches for the study of their behaviour. The progress of this type of analysis will allow for a broader understanding of the collective functions of proteins, which could provide an additional insight on their evolution.